PROCESSING ALUMINA- AND SILICA- BEARING WASTES

By

Viktor L. Rayzman,
Sc.D.

Santa Monica

California

2014

PREFACE

This book summarizes the innovations introduced by the author and his co-authors in theory and practice of extracting alumina and silica from industrial waste and raw material, such as coal ash, red mud, nepheline, and low-grade bauxite. Despite the huge world's reserves of high-quality bauxite a lot of countries involving USA, UK, Japan, Russia, Germany and Canada lack sufficient amount of that resource. Therefore this book will be helpful for all possessors, researchers, designers, operators, and others concerning aluminum and alumina production, coal and shale extraction and power generation at many "nonbauxite" regions. The important scientific result of the explorations embodied in this book lies in development of concepts of the advanced alkaline processing aluminum-containing refuse into alumina and chemicals. The main advantage of the developed technology is "new life of coal" as a term to denote combining use of both combustible and mineral parts of the alumina-bearing coal with simultaneous production of power, liquid and gaseous fuel (from the combustible components), and aluminum, alumina and other inorganical chemicals (from the mineral constituents), involving improvement of environmental conditions compared to the current coal-using plants. In such manner, the countries that don't dispose of bauxite, oil, and gas deposits, could overcome dependence from the producers of the above-mentioned resources using their own solid fuel reserves.

The author expresses his deep gratitude and appreciation to Mr. Stan Vishnevsky for his friendly support and qualified consultation.

INTRODUCTION

The majority of aluminum produced today is recovered from low-silicon bauxite. Good grade bauxite deposits are located at the Earth's torrid zone. Processing bauxite into aluminum is carried out in two steps. In the first stage, aluminum oxide called alumina is extracted from rock on alumina refinery by the Bayer method, then alumina is processed into metal on aluminum smelter by electrolysis [1].

The Bayer process is founded on extracting alumina Al_2O_3 from bauxite by digestion with a reclaimed caustic solution. The following chemical reaction takes place: $Al_2O_3.nH_2O_{(s)} + 2NaOH_{(aq)} = NaAlO_{2(aq)} + (n+1)H_2O_{(L)}$

       alumina hydrate   caustic soln.   aluminate solution

where n is equal to 1 for boehmite and diaspore and 3 for gibbsite.

Insoluble remainder as "red mud" is separated from the obtained aluminate solution by decantation, washing and filtration. Silicon inherent in bauxite as quartz and kaolinite converts to solid sodium hydroaluminosilicate (SHAS) $Na_2O.Al_2O_3.2SiO_2.H_2O_{(s)}$ by the reaction

$2SiO_{2(s)} + 2NaAlO_{2(aq)} + H_2O_{(L)} = Na_2O.Al_2O_3.2SiO_2.H_2O_{(s)}$

SHAS is incorporated into red mud and its amount correlates with alumina loss. The formed aluminate solution supersaturated with AlO_2^- anions is subjected to the precipitation procedure carried out in accordance with the chemical reaction that is the reverse of the digestion procedure as follows:

$NaAlO_{2(aq)} + 4H_2O_{(L)} = Al_2O_3.3H_2O_{(s)} + 2NaOH_{(aq)}$

Most of the alumina hydrate gradually crystallizes out, upon inoculation with $Al(OH)_3$ taken from a previous precipitation stage, and the freshly formed aluminum hydroxide is in sequence washed and calcined to the alumina. The depleted solution and wash water are concentrated and used for another digestion. The Bayer method flow chart is shown in Figure 1.1.

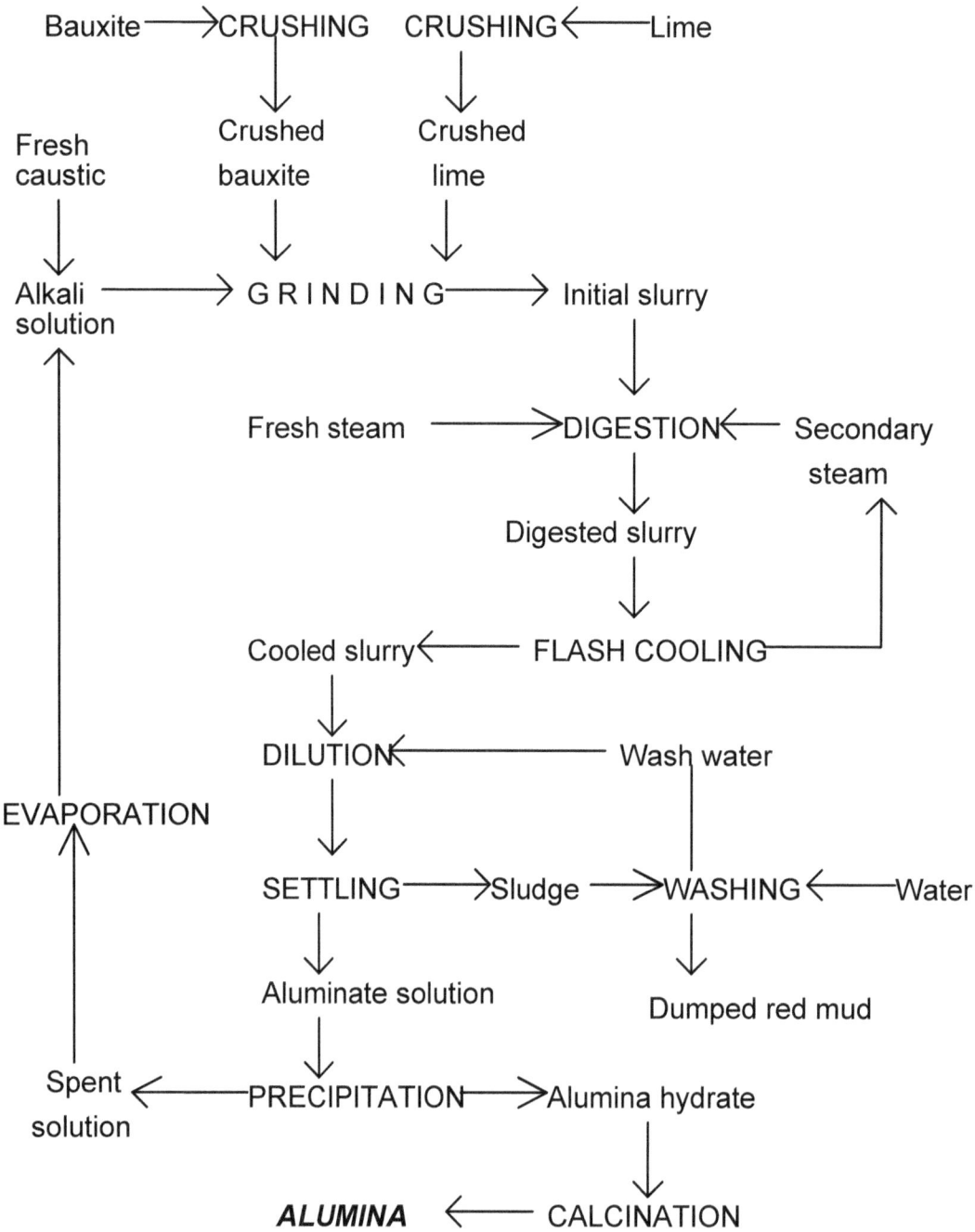

Figure 1.1. Flow chart of the conventional Bayer process

The obtained alumina is subjected to electrolysis in cryolite-alumina melt

4

producing aluminum.

Limited occurrence of the high-quality bauxite impedes the further increasing aluminum industry capacity. To avoid a great outlays on the ore transportation, alumina refineries generally are erected close to bauxite minings. The obtained alumina is delivered to aluminum smelters which are located in the regions of available inexpensive electric power. Different location of mining-refinery units and smelters causes additional cost for shipping alumina. Whereas many countries involving USA, Britain, Germany, Japan, Canada and Russia dispose of huge resources of high-silicon alumina-bearing raw material and waste. Silicon-rich bauxite, coal and shale ash, red mud and nepheline dominate among these sources. However, in this case, the Bayer process cannot be applied because of significant silicon amount presence. The many and varied methods have been developed for extracting alumina from raw material and waste of different composition [2]. These methods will be surveyed in the associated chapters of this book. Based on the accomplished analysis a body of new developed processes has been advanced to notice of experts in metallurgy, chemistry and power engineering areas.

The book contains four chapters. The first chapter is devoted to extracting alumina and silica-bearing constituents from bauxite and red mud. New methods for processing coal ash into alumina and byproducts are embodied in the second chapter. The third one comprises description of advanced methods for utilization of aluminate and silicate solutions obtained in the course of alumina production. New prospective avenues of integrating technology for processing alumina and silica-bearing both coal and raw material are collected in the fourth chapter.

References:

1."Production of Aluminum and Alumina," ed. By A.R. Burkin, John Wiley&Sons, New York, 1987.-241 p.

2.L.P. Ni, V.L. Rayzman, and O.B. Khalyapina, "Alumina Production. Reference Book," Academy & Ministry of Sciences of Republic of Kazakhstan , Almaty, Kazakhstan, 1988.-356 p. (in Russian).

Chapter 1. DEVELOPMENT OF ALKALINE METHODS FOR PROCESSING RED MUD AND HIGH-SILICA BAUXITE

1.1.Chemical Beneficiation of Red Mud

Red mud as a residue resulting from bauxite digestion contains a relatively high amount of less common and rare-earth elements including scandium, which contents is in the 100g/t area. Chemical beneficiation in combination with the Bayer process and, sometimes, with reducing smelting offers a means of obtaining concentrates with scandium percentages within 200-900g/t. Coincindentally with the scandium recovery, an iron-rich product results and both alumina and caustic losses are significantly decreased. The addition of minor amounts of scandium, within 1%, to aluminum and its alloys has a beneficial effect on structure and properties of the products obtained. The positive impact of scandium is more significant than that of other transition metals such as zirconium and titanium. The similar improvements of quality due to scandium admixture extend to Al-Mg alloys [1]. The scandium influence n the properties of Al-Sc stripes is shown in Table 1.1.1 [2].

Table 1.1.1. Properties of the aluminum-scandium strips (lengthwise)

Scandium content (wt%)	0	0.14	0.23	0.36	0.40	0.56
Temporary tensile strength (MPa)	85	98	178	234	246	292
Yield stress (MPa)	67	78	157	216	248	278
Specific elongation (%)	41	39	18	15	16	15

Scandium Production, Value and Use. The extremely low scandium percentage in the available ores causes high prices of the metal and its compounds. For instance, the cost of Sc_2O_3 [1260-08-01] of 99.9% pure was $10,000/kg [3]. World scandium output in 1985 annual (outside the USSR) was appraised as 100 kg. There are indications that 1 kg of metallic scandium cost between $12,000 and $15,000 [4]. Scandium is used for manufacturing low induction ferrites for computers [5]. It is also used in the manufacture of military laser technology [6]. But generally, scandium is of prime interest for rocket building, the aerospace industry and austronautics due to its high melting point which is 2.5 times that of aluminum but the same specific weight [1]. For this purpose, a master alloy (Al 98%, Sc 2%) has been developed [7]. These outlined potential new application for scandium- bearing compounds in automotive and sporting products. Of note is the feedstock

6

together with its complicated scandium recovery process. The ores containing 105g/t Sc are accepted source for scandium production [8]. A more scandium-rich feedstock may be obtained through the chemical beneficiation of red mud dumped by alumina refineries.

Testing the Theory. Throughout the Bayer process (BP), 95-100% of the scandium contained in bauxite remains in red mud [9]. The percentages of scandium and some rare earth elements in various red mud samples are collected in Table 1.1.2 [10].

Table 1.1.2. Content of rare earth element in some red mud samples (g/t)

Company/ Plant	Bauxite source	Sc	Y	Ce	La
Kaiser Aluminum/ Baton Rouge	North Jamaica	101.2	759.7	684	1540
Reynolds Metals/ Corpus Christi	Jamaica (75%) Haiti (25%)	106.6	485.5	591	1508
Alcoa/ Point-of-Comfort	Guinea (60%) Dom. Republic (25%) Surinam (15%)	76.9	293.3	319	651

As is evident from this Table, a conventional red mud can be equated, in most cases, with the ore mentioned in the References [7,8]. But red mud has ponderable advantages over the above-mentioned raw materials. The development and application of a chemical beneficiation method (CBM) has shown to increase significantly the scandium content. Moreover, the combination of BP and CBM has been shown offer an increase in Al_2O_3 yield together with a lowering of caustic losses. A flow chart of the integrated CBM-BP technology supplemented by the material balance with reference to a sample of Australian bauxite is shown in Figure 1.1.1.

BAUXITE-1969.9: SiO_2-95.2;Al_2O_3-1002.0;Fe_2O_3- 287.9; TiO_2-57.2;CaO-1.6;Zr-2.2;S-0.2;Sc-0.069;Y-0.0591;La-0.0197

$\downarrow$

PREPARATION, DIGESTION, FLASHING, SETTLING

$\downarrow$

RED MUD-1019.52(100%):Na_2O 114.4(11.2%); SiO_2.118.9(18.5%); Fe_2O_3 287.9(28%); TiO_2(5.6%);CaO 1.6(0.2%); Zr 2.2 (2,157.9g/t); Sc 0.069(67.8g/t);Y 0.0591(58.1g/t); La 0.0197(19.4g/t)

$\downarrow$

CHEMICAL BENEFICIATION

$\downarrow$

SLIME-241.3(100%):Na_2O 2.0(0.8%); SiO_2 2.0(0.8%); Al_2O_3 2.0(0.8%); Fe_2O_3 171.6(71.3%); TiO_2 5.6(2.3%); CaO 1.6(0.7%); Zr 2.2(0.9%); Sc 0.069(286g/t); Y 0.0591(245g/t); La 0.0197(82g/t)

$\downarrow$

REDUCING SMELTING

$\downarrow$

SLAG-69.9(100%):SiO_2 1.6(2.3%); Al_2O_3 2.0(2.9%); Fe_2O_3 0.6(0.9%); TiO_2 5.6(8.0%); CaO 1.6(2.3%); Zr 2.2(3.1%); Sc 0.069(987g/t); Y 0.0591(845g/1); La 0.0197(281g/t)

PIG IRON-119.9(100%): Fe 119.7(99.83%); Si 0.2(0.17%)

Figure 1.1.1. An example of materials distribution of the in-series method involving Bayer process, chemical beneficiation and reducing smelting as applied to Australian bauxite (kg/t Al_2O_3)

Both the flow chart and balance presented are finished to the reducing smelting stage (RSS), although the process may be restricted to CBM, only, giving the concentrate containing Sc 286 g/t, Zr 0.9%, as well Fe_2O_3 71.3%. Simultaneously, an additional recovery of Al_2O_3 and Na_2O from red mud takes place. Scandium content in slag after RSS is increased to 987g/t. The results of both bench-scale and pilot-plant tests are shown in Table 1.1.3.

Table 1.1.3.Composition of initial bauxites and products obtained in the course of scandium concentration

Process used	Bauxite origin & product obtained	SiO$_2$ (%)	Al$_2$O$_3$ (%)	Fe$_2$O$_3$ (%)	TiO$_2$ (%)	Na$_2$O (%)	Sc (g/t)	Yield (%): A-Al$_2$O$_3$ N-Na$_2$O S –SiO$_2$
Bayer	*Australia*	9.9	50.9	14.6	2.9	1.6	35	
	Red mud	18.5	18.5	28.0	5.6	11.2	68	A 81.1
CB*	Slime	0.8	0.8	21.3	23.7	0.8	286	A 98.9
RS**	Slag	2.3	2.9	0.9	8.0	0	987	
	Hungary	7.1	51.4	23.7	2.5	0.1	70	
CB*	Slime	1.1	2.8	75.4	11.0	0.7	200	A 98.3
	Jamaica	0.2	46.7	18.7	2.4	0.5	<50	
HTBP***	Red mud	1.6	4.7	64.9	8.3	1.4	230	A 97.2
	Kazakhstan	7.4	42.8	18.3	1.4	0.2	33	
Bayer	Red mud	20.4	21.7	32.4	3.3	10.2	55	A 71.4
CB*	Slime	6.4	6.0	43.0	13.2	3.3	110	A 93.0 N 93.1
RS**	Slag	15.0	7.3	2.2	29.8	0	420	
	Russia	17.6	51.9	8.0	3.0	0.2	78	
1)Bayer	Red mud	22.5	26.3	11.1	3.4	16.1	110	A 60.0
CB*	Slime	5.8	5.9	28.7	22.0	3.0	700	A 97.0 N 97.6
2)R-CB****	Calcine	9.7	72.4	6.0	3.8	1.5	120	S 62.3
Bayer	Red mud	18.4	20.3	13.2	8.0	11.4	370	A 86.8
3)CB*	Slime	8.0	4.9	44.8	16.8	4.7	480	A 98.4

NB: CB* - Chemical Beneficiation; RS **- Reducing Smelting; HTBP*** - High Temperature Bayer Process ; R-CB**** - In-series Roasting followed by CB

These data were obtained by processing bauxites and red muds produced in various parts of the world as follows: Hungary [11], Russia [12], Kazakhstan [13], and Jamaica [14]. As can be seen from the tabular data, the combination of BP, CBM and RSS is subjected to variation. Hungarian bauxite was processed by CBM, directly. As a result, scandium content in concentrate obtained ranged up to 200 g/t and alumina extraction yield reached 98.3%. The Bayer process was modified and pilot-plant tested with Jamaican bauxite [14,15]. By increasing the temperature to 285-300°C, red mud containing 230g Sc/t was obtained with alumina yield increased by 5% in comparison with the routine BP. A complete cycle was tested on Kazakhstan Red October bauxite low in scandium at 55g/t [13]. This feedstock was processed by CBM and the resulting concentrate contained 110g Sc/t.

Simultaneously, 93% Al_2O_3 and 93.1% Na_2O were extracted. Finally, slag containing 420g Sc/t was produced by RSS. Thus, the scandium content was increased 12.7 times compared with the initial bauxite. Three avenues have been tested for introduction of combined processes on Russian North Onega bauxite [12]. The most scandium-rich product was obtained by in-line BP-CBM with extraordinary yield of both Al_2O_3 97% and Na_2O 97.6%.

Advantages of the Chemical Beneficiation Method. Methods already exist for extracting scandium from red mud. These processes involve RSS followed by slag sulfurisation [1] or direct red mud treatment with hydrochloric acid using organical extractants [16]. Problems with these methods are, on the one hand, an inadequate scandium content at about 100 g/t, and, on the other, an intolerable percentage of both Al_2O_3 and Na_2O in red mud. These disadvantages cause excessive acids and power consumption, as well low output of scandium and, consequently, its high cost. From this standpoint, the proposed CBM provides a depletion of red mud with aluminum and sodium and, on the contrary, a gain in scandium content to 200 g/t and higher. These circumstances together with demands made on aluminum-scandium alloys appear favorable to large producers operating both alumina refineries and aluminum smelters. The combination of BP and CBM presents the opportunity to increase alumina yield, decrease caustic losses and takes a lead in the special alloys market.

References:

1.B.G. Korshunov, A. M. Reznik, S. A. Semyonov, "Scandium", Metallurgiya:Moscow,1987,184p.

2.A.V. Boldyrev, "Mechanical and Technological Properties of Metals", 2nd ed., Metallurgiya: Moscow, 1987, 208 p.

3.Gondyi Guo, Yuli Chen, "Solvent Extraction of Scandium from Wolframite Residue", J.Metals, 1988, 40(7), p. 28-31.

4.M. Mosin, Report from the Closed Zone", Pravda (Russia), November 10, 1989, 2.

5. B.I. Kogan, "Rare Metals. Conditions and Properties", Nauka: Tsvetmetinformatsiya: Moscow, 1974, 356 p.

6."New Deposits of Scandium-Containing Raw Material in Norway, Tsvetmetinformatsiya: Moscow, 1988, p. 7-8.

7."Ashurst Makes Progress with Scandium Alloy Development", Aluminum Today, December 1997, p. 18-20.

8.R. Woodward, "Where Next Wrought Aluminum Alloys?", Aluminum Today, December 1997, p. 21-23.

9.V.A. Derevyankin, T.I. Porotnikova, E.K. Kocherova, et al., "Behavior of Scandium and Lanthanum on Aluminum Production from Bauxites", Izv. Vyssh. Uchebn. Zaved., Tsvetn. Met., 1981, 4, p. 86-89.

10.S. Vijayan, A.J. Mehuk, D. Singh, "Recovery of Scandium from Jamaican Red Mud", Atomic Energy of Canada Ltd.: Pinawa, Manitoba, October 1986.

11.V.L. Rayzman, "Investigation of Process for Hydrothermal Dressing of Red Muds", Research Report/LOZ VAMI: Leningrad, 1990, 32 p.

12.V.L. Rayzman, R.A. Abdulvaliev, L.P. Ni, et al., "Decomposition of Red Mud", Transactions of VAMI: Leningrad, 1990, p. 7-14.

13.V.L. Rayzman, V.I. Pevzner, et al., "Investigation of Method for Manufacture of Rare Metals Concentrate by the Technique of Limeless Hydrochemical Digestion of Red Mud Produced from High Ferrous Bauxite", Research Report No 01900048600/ VAMI: Leningrad, 1990, 85 p.

14. V.L. Rayzman, V.I. Pevzner, L.P. Ni, R.A. Abdulvaliev, et al., "Development of Process for Manufacture of Product Enriched with Scandium", Research Report/ LOZ VAMI, IMIO AN KazSSR: Leningrad, Alma-Ata, 1986, 54 p.

15.N.A. Kaluzhskii, V.L. Rayzman, V.V. Meshin, et al., "Production of Scandium-Containing Concentrate from Low Silicon Bauxite under Pilot Plant Conditions", Tsvetn. Met., 1990, 7, p. 73-74.

16. S.A. Semyonov, A.M. Reznik, L.D. Yurchenko, "Extraction of Scandium During the Complex Processing of Various Raw Materials", Tsvetn. Met., 1983, 12, p. 43-47

1.2. Consecutive Recovering Silica and Alumina from High-Silica Bauxite with Different Fe_2O_3 Percentages

Due to increasing demand by alumina refineries, bauxite mining capacity is increasing at the rate of 2 million tonnes per year [1]. Low-grade, high-silica deposits, which are located throughout the world, could be used to meet the growing demand. Examples of associated bauxite compositions are shown in Table 1.2.1.

Table 1.2.1. Composition of low-grade bauxite of some world deposits (wt %)

Country/Location	Al_2O_3	SiO_2	Fe_2O_3	TiO_2	LOI	Main Al_2O_3 and SiO_2-bearing minerals
United States: Arkansas	50–55	11–13	2–6	3–4	28–30	gibbsite, kaolinite, quartz[2]
Russia: Arkhangelsk District	51–56	16–20	6–9	2–2.8	16–17	boehmite, gibbsite, kaolinite [2]
Komi Republic	45–50	5–12	25–30	2–5	12–16	boehmite, shamozite[2]
Kazakhstan	41-46	10-13	15–16	1.8–2.2	23–24	gibbsite, kaolinite [2]
Australia: Weipa	54–55. 5	5–6	11–14	—	24–26	gibbsite, boehmite, kaolinite [3]
China	68.7	9.07	5.22	3.32	13.93	diaspore, illite [4]

Silica-bearing minerals of bauxite interact with aluminate solution according to Equation 1

$$2SiO_{2(s)} + 2NaAl(OH)_{4(aq)} + 2H_2O_{(L)} = 2NaOH.Al_2O_3.2SiO_2.H_2O_{(s)} \quad (1)$$

The solid residue of the formed sodium hydroaluminosilicate has one NaOH mole and 0.5 Al_2O_3 mole per one SiO_2 mole. As a result, 0.666 kg NaOH and 0.85 kg Al_2O_3 per 1 kg of silica are lost irrevocably. As shown in Table 1.2.1, low-grade bauxite contains 5-20% silica, consequently, the loss of useful constituents may reach 132 kg NaOH and 170 kg Al_2O_3 per tonne of the bauxite. Therefore, the partial digestion of silica from bauxite before converting it into alumina leads to a decrease in precipitation of aluminosilicate and an increase of alumina yield in the Bayer process. This method for improving bauxite quality is called thermochemical alkaline conditioning (TCAC) and it involves the roasting of bauxite and alkaline treatment of calcined material.

Conversion of Minerals in Bauxite Conditioning. Consider the two most important silicon-containing minerals of bauxite: kaolinite ($Al_2O_3.2SiO_2.2H_2O$) and quartz (SiO_2) [5]. Quartz converts into tridimite at a temperature approaching 870°C and then transfers into crystobalite. Both of the formed compositions have had the same formula (SiO_2) and can be dissolved in alkali solution by reaction:

$$SiO_{2(s)} + 2NaOH_{(aq)} = NaH_2SiO_{4(aq)} \quad (2)$$

At 800-1,000°C, kaolinite converts into difficult-to-decompose mullite ($3Al_2O_3.2SiO_2$) and crystobalite. Heating alumina-forming hydrates such as gibbsite ($Al_2O_3.3H_2O$), boehmite, and diaspore (both $Al_2O_3.H_2O$) at 900-1,000°C leads to the formation of γ-Al_2O_3 that can be dissolved in caustic solution at 240-280°C. Crystobalite is separated from mullite and γ-alumina by treatment of the calcined bauxite with caustic solution containing 100-200 gNaOH/L at 85-95°C. The obtained solids, called alumina concentrate, are processed into alumina by the Bayer method. As to a silicate solution, it is preliminarily separated then desilicated. The in-series TCAC-Bayer process, shown schematically in Figure 1.2.1, was tested at the alumina-aluminum pilot plant of the Russian National Aluminium-Magnesium Institute (VAMI) [6].

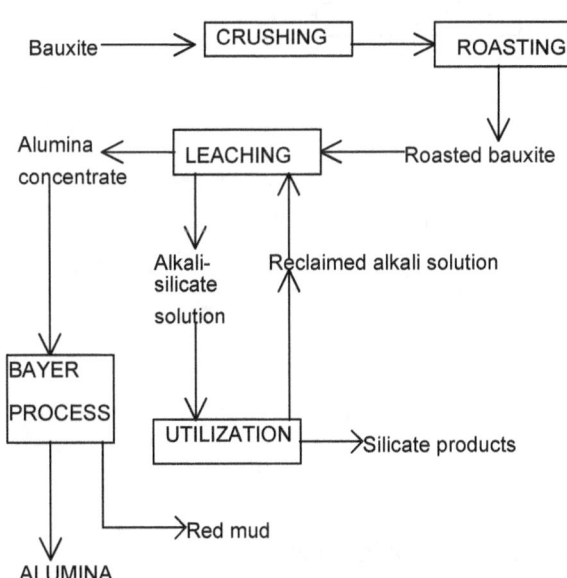

Figure 1.2.1. A flow chart of the combined TCAC-Bayer method implemented at VAMI plant

Processing Bauxite. Three low-grade bauxite lots that vary in Fe_2O_3 content were processed: low-iron (Fe_2O_3 7.2%) Arkhangelsk ore [7], high-iron (Fe_2O_3 27.5%) Komi bauxite [8], and medium-iron (Fe_2O_3 17.6%) Kazakhstan rock [9]. The raw material was put through a series of crushing mills to achieve a piece size of -20 mm and then roasted at temperatures of 950-1,000ºC in the rotary kiln (Figure 1.2.2).

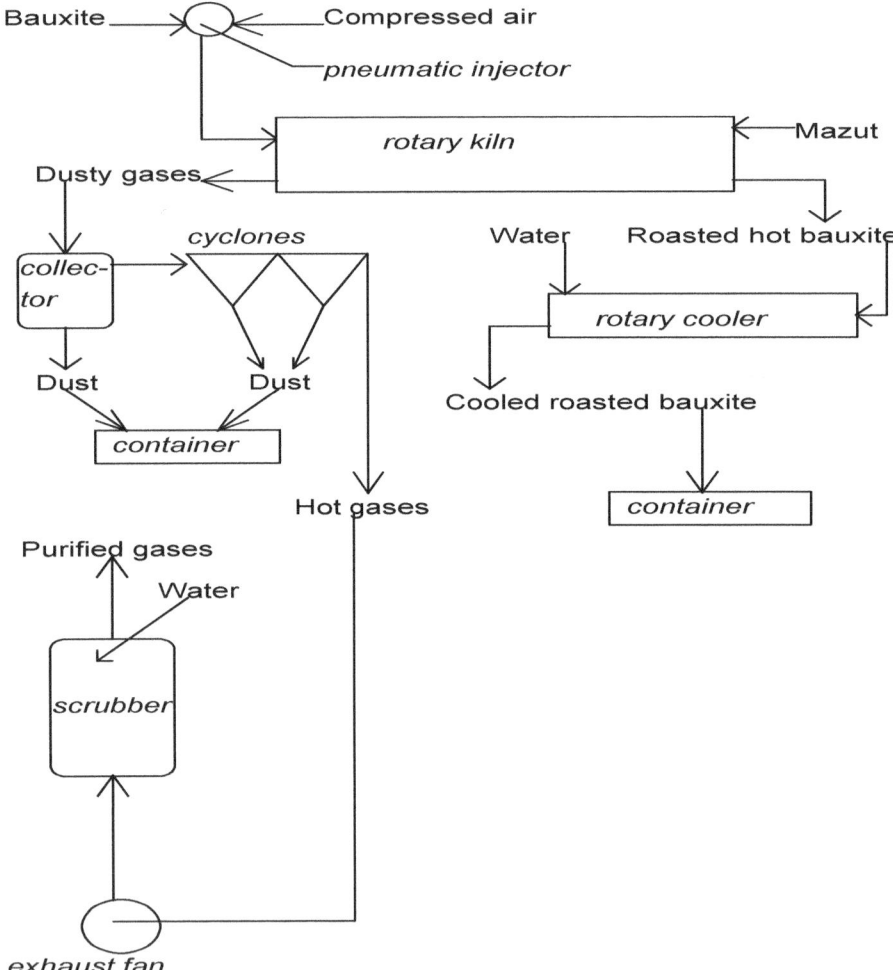

Figure 1.2.2. A flow chart of the unit destined for bauxite roasting

Bauxite was delivered to the kiln by pneumatic injector. The residence time of bauxite in the roasting zone was 30-40 min. The clinker was passed through a rotary cooler with externally refluxing water, and the obtained calcine was collected in a container. Flue gases were exhausted by a fan and prepurified in a dust collector and cyclones. Then, in the case of the Arkhangelsk bauxite, the roasted material was reground to -6 mm pieces and classified into two fractions: -6+2 mm and -2 mm. The first fraction was desilicated (leached) by percolation in a conveyer-type extractor. A percolator (Figure 1.2.3) consists of a chain conveyer equipped with netted bottom containers.

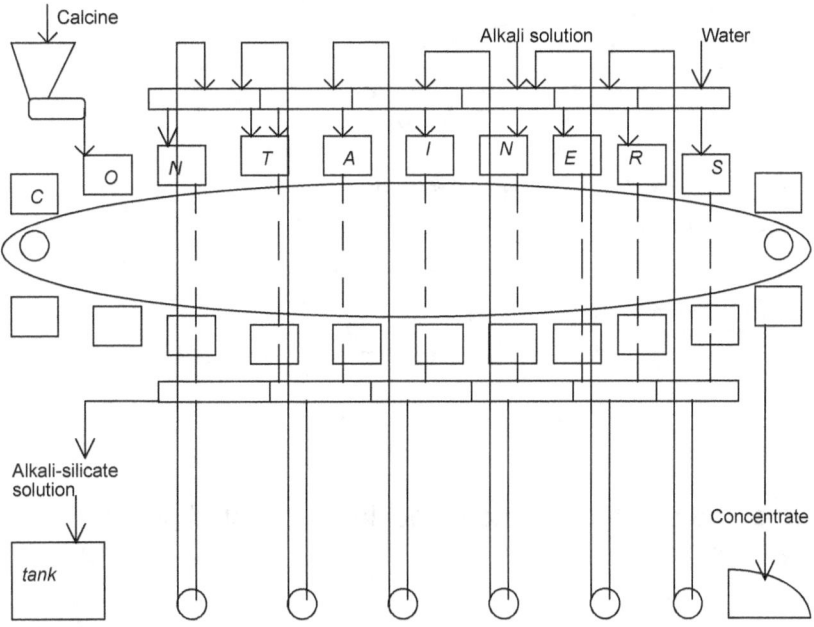

Figure 1.2.3. An outline of the percolator for leaching of roasted coarsely crushed bauxite with a high silicon percentage

The percolation leaching procedure was carried out by filtration of solution through a leached material bed stacked on the bottom screen. The countercurrent principle of calcine and solution movement was employed. Hot water was fed into a last percolation zone to wash the obtained concentrate. The initial alkali solution was directed to the next zone to leach the first portion of roasted bauxite and then it passed through all calcine-containing zones until the final discharge of alkali-silicate solution from the primary zone. The rotation of the percolator conveyer provided a total leaching and rinsing residence time of material in the extraction zone of six hours. The solution temperature throughout the zones was kept within 85-95°C. Alumina loss into solution ranged from 1% to 2%. The second fraction was leached by agitation. The other two mentioned ores were ground after roasting to a size of -2mm and also leached by agitating. The unit shown schematically in Figure 1.2.4 is designed for the agitation leaching procedure.

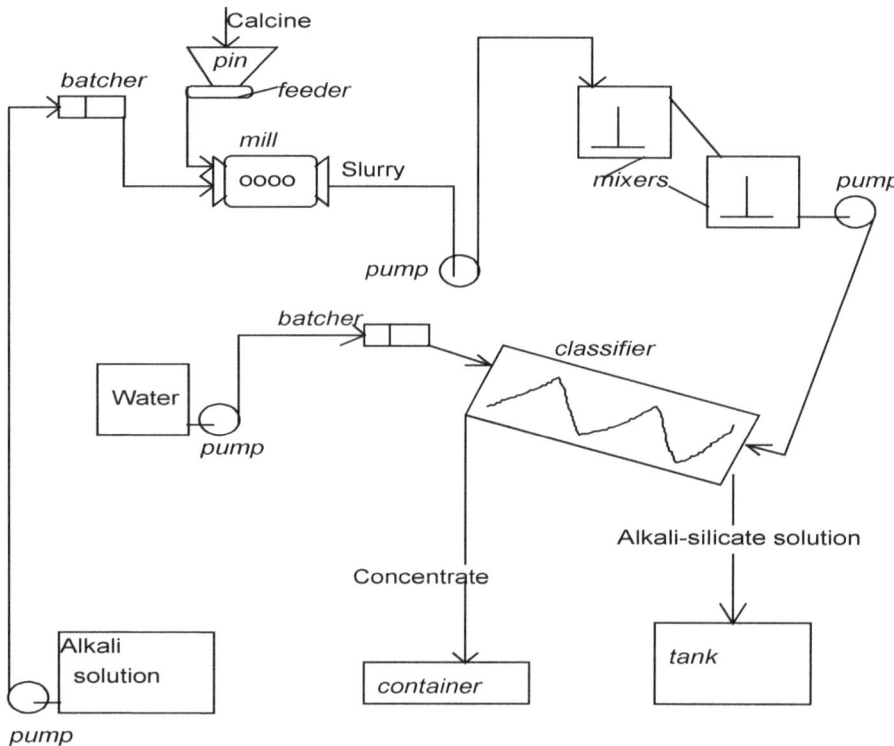

Figure 1.2.4. A flow chart of the unit for agitation leaching of roasted finely crushed high in silicon bauxite

Finely ground (-2 mm) calcined bauxite was dosed from a bunker by feeder to a rod mill (L = 3 m, D = 1 m) revolving at 32 rpm. Initially, the heated alkali solution at 90-95°C was fed to the mill with pump and batcher. The alkali-bauxite slurry with a liquid-to-solid ratio of 5:1 was delivered into two in-line stirrers by pumps. A pump transferred agitated slurry to a spiral classifier where the solid residue was separated from the obtained solution. The leached residue (i.e., alumina concentrate) was rinsed at the classifier rear zone by hot water and collected into containers. Overflow (alkali-silicate solution) was discharged to a stirrer designed for desilication. The obtained liquor was subjected to final filtration by a press filter. Filtration productivity was equal to 780 L/(m2.hr). The alumina concentrate obtained from percolation and agitation units was processed by the well-known Bayer method with high temperature digestion. The arrangement of corresponding facilities is schematically depicted in Figure 1.2.5.

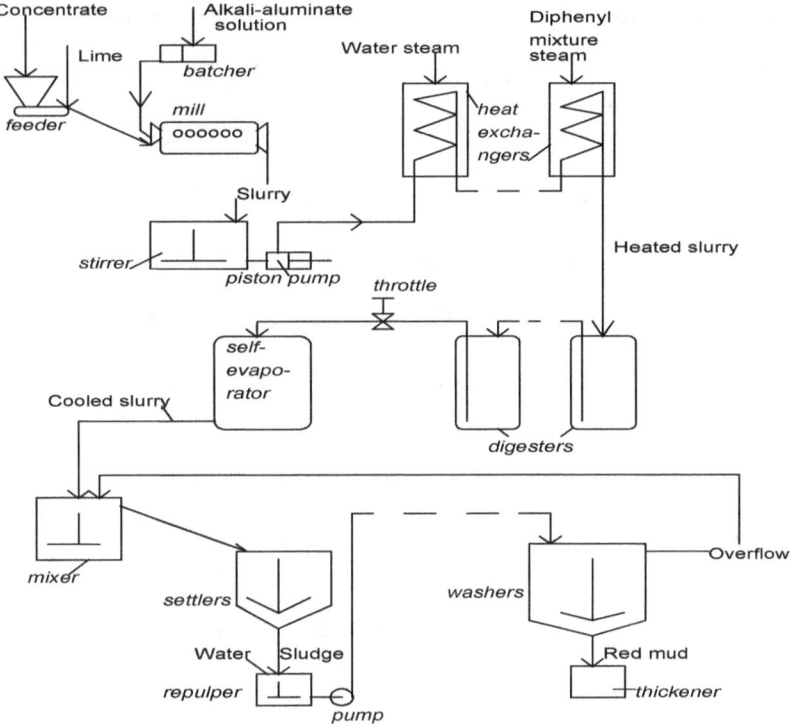

Figure 1.2.5. A flow chart of the unit for processing the concentrate obtained by the Bayer method

Concentrate along with added lime at the rate of 3% of the feedstock input was mixed with alkali-aluminate solution and ground in a double-chamber mill. The obtained slurry was pumped to the bank of digesters with steam heat exchangers arranged in sequence. The first heating line used water steam, but the second one was heated by steam of diphenyl mixture generated by a special boiler. After going through digesters, the slurry was discharged through a throttle and self-separator into a mixer for dilution with the first process water. The dissolved slurry was settled and separated into sludge and overflow. The sludge was rinsed with water to produce dump red mud and reclaimed process water. The overflow, also called aluminate solution, can be processed into alumina or sodium aluminate (hydroaluminate) [10]. Unlike the conventional Bayer method, digestion temperature for recovery of alumina was increased from 200-220°C to 260°C due to the presence of difficult-to-decompose γ-alumina and mullite. The residence time of digestion was varied from one hour (for high-iron bauxite) to two hours (for low-iron ore). In any case, alumina yield was 19-20% higher than the same parameter for crude feedstock. The results (Table 1.2.2) show that, specifically, 57.3-60.7% of silica was extracted into alkali-silicate solution and 86.7-89.5% of alumina was recovered by the Bayer method.

Table 1.2.2. Composition of three bauxite lots and products obtained by their processing with the combined TCAC-Bayer method

	Lot # 1			Lot # 2			Lot # 3		
Solid material constituents	Al_2O_3	SiO_2	Fe_2O_3	Al_2O_3	SiO_2	Fe_2O_3	Al_2O_3	SiO_2	Fe_2O_3
Raw bauxite (wt.%)	53.6	18.2	7.2	46.9	9.4	27.5	51.0	9.21	17.6
Roasted bauxite (wt.%)	59.3	21.1	9.5	51.3	12.3	31.4	56.5	12.7	24.3
Concentrate (wt.%)	67.4	11.1	11.7	57.2	4.9	30.4	61.2	5.4	26.3
Red mud (wt.%)	13.9	11.1	23.0	9.3	10.9	42.1	12.1	11.0	39.1
Liquid material concentration (g/l)	NaOH	SiO_2	Al_2O_3	NaOH	SiO_2	Al_2O_3	NaOH	SiO_2	Al_2O_3
Alkali-silicate soln.	128.1	24.7	-	200.0	22.5	-	99.0	27.7	-
Yield (%)		57.3			58.9			60.7	
Aluminate soln.	347.0	-	267.0	333.0	-	257.0	316.0	-	238.0
Yield (%)			89.5			88.3			86.0

The washed red mud can be, in its turn, subjected to further chemical beneficiation to produce additional alumina and caustic, as well as iron and rare metals-bearing concentrate containing about 70% Fe_2O_3, 1% zirconium, 300 g of scandium per tonne, 250 g of yttrium per tonne, and 80-85 g of lanthanum per tonne (see 1.1).

Utilization of Alkali-Silicate Solution. The bulk of such liquid intermediate was subjected to desilication by lime treatment. Calcium hydrometasilicate ($CaO.SiO_2.H_2O$) is precipitated by the reaction:

$$Na_2H_2SiO_{4(aq)} + CaO_{(s)} + H_2O_{(L)} = CaO.SiO_2.H_2O_{(s)} + 2NaOH_{(aq)} \quad (3)$$

Regenerated caustic solution can be used to leach roasted bauxite. The procedure of desilication was carried out by conventional chemical processing unit operations, including stirrers equipped with heat exchangers, pumps, tanks, and filters. The residence time of the desilication procedure at 90-95°C and a molar ratio of CaO: SiO_2 = 1.1:1 is three hours. A portion of alkali-silicate solution was processed into granulated sodium silicate. Such chemical compound can be used for manufacturing zeolites, construction materials, fabric bleach, and drilling slurry. The solution was strengthened to concentrations of 360 g NaOH/L and 230 g SiO_2/L and then dried and granulated in a fluidized bed furnace shown schematically in Figure 1.2.6 [6].

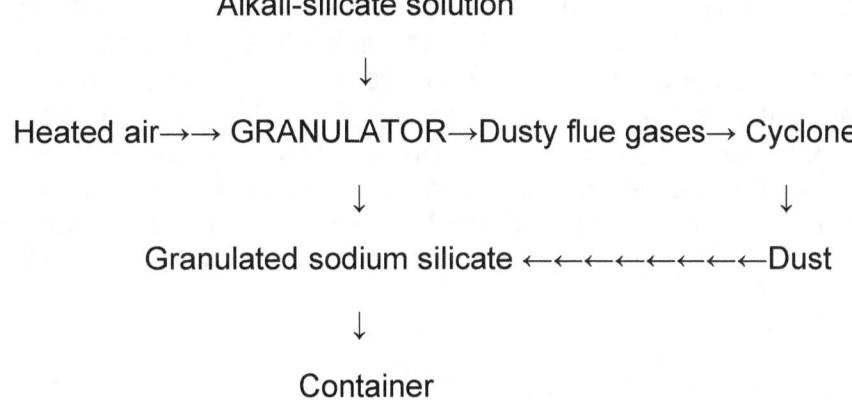

Alkali-silicate solution

↓

Heated air→→ GRANULATOR→Dusty flue gases→ Cyclone

↓   ↓

Granulated sodium silicate ←←←←←←←←Dust

↓

Container

Figure 1.2.6. An outline of the fluidized bed granulator for utilization of alkali-silicate solutions

The furnace body was shaped as a cylinder narrowing into cone. Solution entered into the fluidized bed from above through a spray nozzle. A perforated hearth was set up at the furnace bottom. There was an unobstructed settling area in the upper part of the apparatus. Air heated at 800-850°C was used as both a heat carrier and fluidizing agent. Dusty fuel gases were cleaned in-line in paired cyclones and then in a wet scrubber. Captured dry dust was unloaded from cyclones and applied as a powdered sodium silicate. Scrubbing wastewater was returned for granulation. Optimal operation parameters were as follows:

1)Temperature: under hearth 800°C, in the fluidized bed 400°C.

2)Velocity: under hearth 6.5 m/s; in the fluidized bed 2.2 m/s; of the sprayed solution 32.2 m/s.

3)Pressure differential in the fluidized bed: 5.5 kPa.

The product unloaded from the granulator had a form of spherical granules with an average particle size of 5.5 mm. Granule water solubility was 99.5-100%. The chemical composition of commodities was:

1)Granules-Na_2O 54.3%, SiO_2 35.4%, ignition loss 1.03%.

2)Dust-Na_2O 53.3%, SiO_2 34.9%, ignition loss 1.8%.

Conclusion. The developed and tested process for manufacturing alumina concentrate and silicate products from high-silica bauxite makes it possible to significantly extend resource reserves for alumina production, not only quantitatively, but geographically as well. It provides a raw material for alumina refineries very close to aluminum smelters and reduces transportation expenses. Moreover, conditioning of bauxite makes it possible to replace costly sintering or Bayer-sintering processes with the less expensive TCAC-Bayer method in various regions of the United States, Australia, Russia, China, and Kazakhstan.

References:

1. W.B. Morrison and R.F. Nunn, "Future Bauxite Supplies to the Alumina Industry," Reports of the TMS Annual Meeting (Poster), San Diego, CA, 1998.

2. V.L. Rayzman et al., "Chemical Conditioning of High Silicon Aluminum-Containing Raw Material," Tsvetmetinformatsiya (in Russian), vol. 3 (1987), pp. 1-60.

3. Processing of Weipa Bauxite (Prospect) (Melbourne, Australia: Comalco Company, 1986).

4. Z. Lun-he, X. Qi-li, and W. Xiang-li, "The Effect of Various Additives on Bayer Digestion of Diasporic Bauxite," Light Metals 1996, ed. Wayne Hale (Warrendale, PA: TMS, 1996), pp.59-63.

5. L.P. Ni, V.L. Rayzman, and O.B. Khalyapina, Alumina Production (in Russian) (Almaty, Kazakhstan: Institute of Metallurgy and Enrichment of Kazakhstan Academy of Sciences, 1998).

6. L.P. Ni and V.L. Rayzman, Combined Methods for Processing Low-Grade Aluminum-Containing Raw Material (in Russian) (Almaty, Kazakhstan: Nauka,1988).

7. V.L. Rayzman et al., "Chemical Conditioning of High-Silicon Bauxite," Development of High Efficient Equipment in Aluminum Production (in Russian) (St. Petersburg, Russia: VAMI, 1985), pp. 48-54.

8. O.A. Dubovikov et al., "Desilication of the Roasted Bauxite by Alkaline Solutions," Nonferrous Metallurgy (in Russian), No 10 (1980), pp. 41-43.

9. L.P. Ni, B.E. Medvedkov, and E.B. Medvedkov, "Processing Low-Grade Bauxite," Transactions of Kazakh Academy of Sciences: Processing Substandard Raw Material into Alumina (in Russian), Alma-Ata (1987), pp. 32-38. 10. V.L. Rayzman et al., "Sodium Aluminate and Hydroaluminate. Production and Application," Rostov-na-Donu University: Rostov, Russia, 1990, 120 p.

Chapter 2. UTILIZATION OF ALUMINA-BEARING COAL ASH

Coal is a major contributor to power and heat generation in the United States. Approximately 55% of the U.S. energy is generated from coal. In 2000, almost one billion metric tons of coal was burned, which generated 120 million metric tons of coal combustion products. Electric utilities burned over 860 million tons of coal and generated almost 100 million tons of the coal combustion products [1]. According to economic forecast coal mining and consumption as fuel will have increased in the future from 300 GW in 2004 to 397 GW in 2025 [2]. Simultaneously, deposits of ash from coal combustion are expected to accumulate extensively. Today about 25 per cent of coal combustion solid waste is used for cement and concrete production. Many millions of tons of aluminum and silicon contained in the coal wastes are annually lost in landfills. Use of coal ash as a raw material for production of chemicals and construction materials can significantly fulfill the U.S. need for these commodities while at the same time decrease dusty pollution. A fairly large number of processes for recovering alumina and byproducts from coal ash have been proposed and developed.

2.1. Review of the Methods Dedicated for Producing Alumina and Byproducts from Coal Ash

Commercial utilization of coal ash is limited to production cement, concrete, structural fills, waste stabilizations, and road-building materials [3,4]. Therefore, development and commercialization of new avenues for full-scale processing ash into metals and other valuable products, by chemical and metallurgical methods are attractive. This line of investigation, called "High-Technology Ash Application" [4], covers a wide range of methods for alkaline, acidic, combined, hydro, and dry metallurgical ash processing. This information and its analysis are essential to properly evaluate and select the optimal parameters for ash treatment according to its properties and composition. Moreover, it is important to create methods that could be competitive with well-established technologies for manufacturing alumina and other products. In the U.S., coal-using power plant operations create a need for considerable landfill space to dispose of coal ash. Most coal-fired waste has to be placed in landfills. Continuing reductions in ash storage capacity are expected. Therefore, addressing the problem of the coal ash utilization by recovering its useful components becomes an important task.

Coal Ash Composition. Table 2.1.1 shows the mineral matter composition of some typical United States coals [5]. A major opportunity for coal ash utilization is provided by the fact that 70% or more of waste generated by coal-burning power plants is silica and alumina. On this basis, it might be possible to produce dozens of million tons of aluminosilicates (e.g., zeolites), alkali aluminates, as well as alumina and silicate chemicals. The last two are of particular interest as commodities.

Table 2.1.1. Ash Characteristics of the typical U.S. coal

	Low-volatile bituminous	High-volatile bituminous				Subbituminous	Lignite	
Index number	1	2	3	4	5	6	7	8
State	WV	OH	WV	IL	UT	WY	TX	ND
Ash (dry basis) (%)	12.3	14.1	10.9	17.4	6.6	6.6	12.8	11.2
			Analysis of ash (wt %)					
SiO_2	60.0	47.3	37.6	47.5	48.0	24.0	41.8	28.4
Al_2O_3	30.0	23.0	20.1	17.9	11.5	20.0	13.6	11.0
TiO_2	1.6	1.0	0.8	0.8	0.6	0.7	1.5	0.4
Fe_2O_3	4.0	22.8	29.3	20.1	7.0	11.0	6.6	14.0
CaO	0.6	1.3	4.2	5.8	25.0	26.0	17.6	18.0
MgO	0.6	0.8	1.2	1.0	4.0	4.0	2.5	5.0
Na_2O	0.5	0.3	0.8	0.4	1.2	0.2	0.6	3.6
K_2O	1.5	2.0	1.6	1.8	0.2	0.5	0.1	0.7
Total	98.8	98.5	95.6	95.3	97.5	86.4	84.3	81.1
Ratio of SiO_2:Al_2O_3	2.0	2.0	1.87	2.65	4.17	1.2	3.07	2.58

The mineral matter includes the following most frequently reported minerals [5]: kaolonite $Al_2O_3.2SiO_2.2H_2O$, illite $K_2O.3Al_2O_3.6SiO_2.2H_2O$, pyrite FeS_2, dolomite $CaCO_3.MgCO_3$, calcite $CaCO_3$, ankerite $2CaCO_3.MgCO_3.FeCO_3$, and quartz SiO_2. Change in coal minerals during combustion in a boiler furnace is shown in Figure 2.1.1 [6].

The slagging and clinkering processes start at 880°C. The main components of the bottom and fly ash are mullite $3Al_2O_3.2SiO_2$, quartz SiO_2, calcite $CaCO_3$, hematite Fe_2O_3, magnetite Fe_3O_4, cristobalite SiO_2, and soda-lime glass [7-9]. Oxidation of sulfur contained in the coal during its combustion causes air pollution. Therefore, coal desulphurization is the first problem which could be solved in order to use fossil fuel as a source for heat and power generation.

Temperature,oC						
2300			VOLATIZATION			
2200			of Al2O3			
2100	VOLATIZATION					
2000						
1900	of SiO2		VOLATIZATION			
1800						
1700						
1600	SLAGGING					
1500			of			
1400						
1300						
1200	and					
1100			ALKALIES			
1000						
900	CLINKERING		ONSET			
800	EVOLUTION		of			
700	of					
600	CO2,SO2,andSO3		SINTERING			
500						
400	OXIDATION CHANGE in MINERAL FORMS					

Figure 2.1.1. Effect of heating on minerals in coal

Coal Desulfurization. It is essential to protect the environment from the sulfur-contaminated gases from coal combustion. Clearly, sulfur or sulfates should be, at best, the final chemical products of coal desulphurization processes. For instance, the Battelle process [10-14] provides the removal of pyritic, organic, and sulfate sulfur, as well as ash from the coal before its combustion. The process involves mixing the coal with an alkali and calcium hydroxide (Figure 2.1.2), heating the resulting mixture to 220-350°C, cooling, and dividing it into a sodium sulfide-containing solution and beneficated coal. The solution is converted by carbonization to sodium carbonate and sodium sulfide that are processed into a marketable sulfur. Treating the soda solution with lime regenerates the alkali solution and forms calcium carbonate which is calcined to the circulating CaO. The Meyers process is based on treating coal with reclaimable aqueous ferric solution followed by the separation of sulfur and $Fe_2(SO_4)_3$ [14-17].

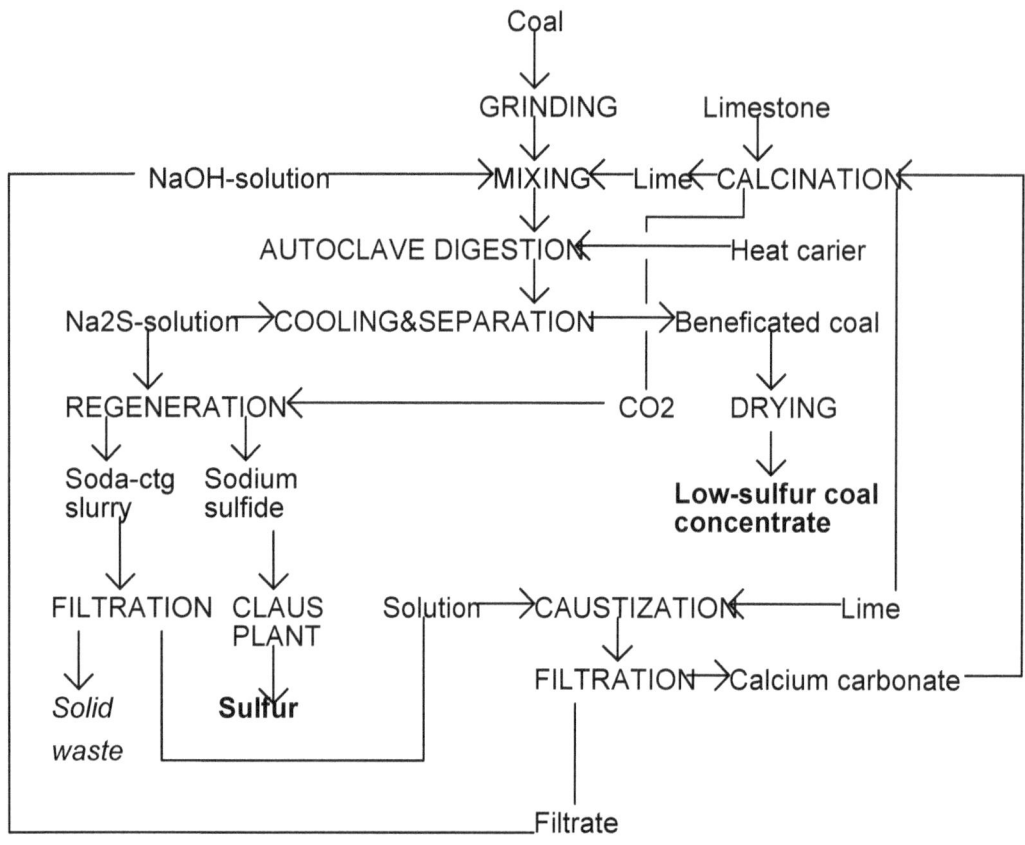

Figure 2.1.2. Flow sheet for the Battelle process

Integrated Technology for Coal Combustion and Ash Sintering. This method was initially developed in the late 1960s. The processes proposed and investigated can be classified in two groups [18-22]. The first is related to alkaline leaching and the second to acidic treatment of the sintered material. In the first technological group, a process was proposed to produce alumina by calcinations of a mixture of aluminum-containing coal and lime or limestone in a $CaO:SiO_2$ ratio of 2:1, a $CaO:Al_2O_3$ ratio of 1.3-1.8:1, and $CaO:Fe_2O_3$ ratio of 1-2:1. The sinter is leached with soda solution, the slurry is filtered, and the sludge is washed with water and used for cement production. Alumina is obtained by desilication and carbonization of the filtered aluminate solution and subsequent calcination of the precipitated aluminum hydroxide (Figure 2.1.3).

24

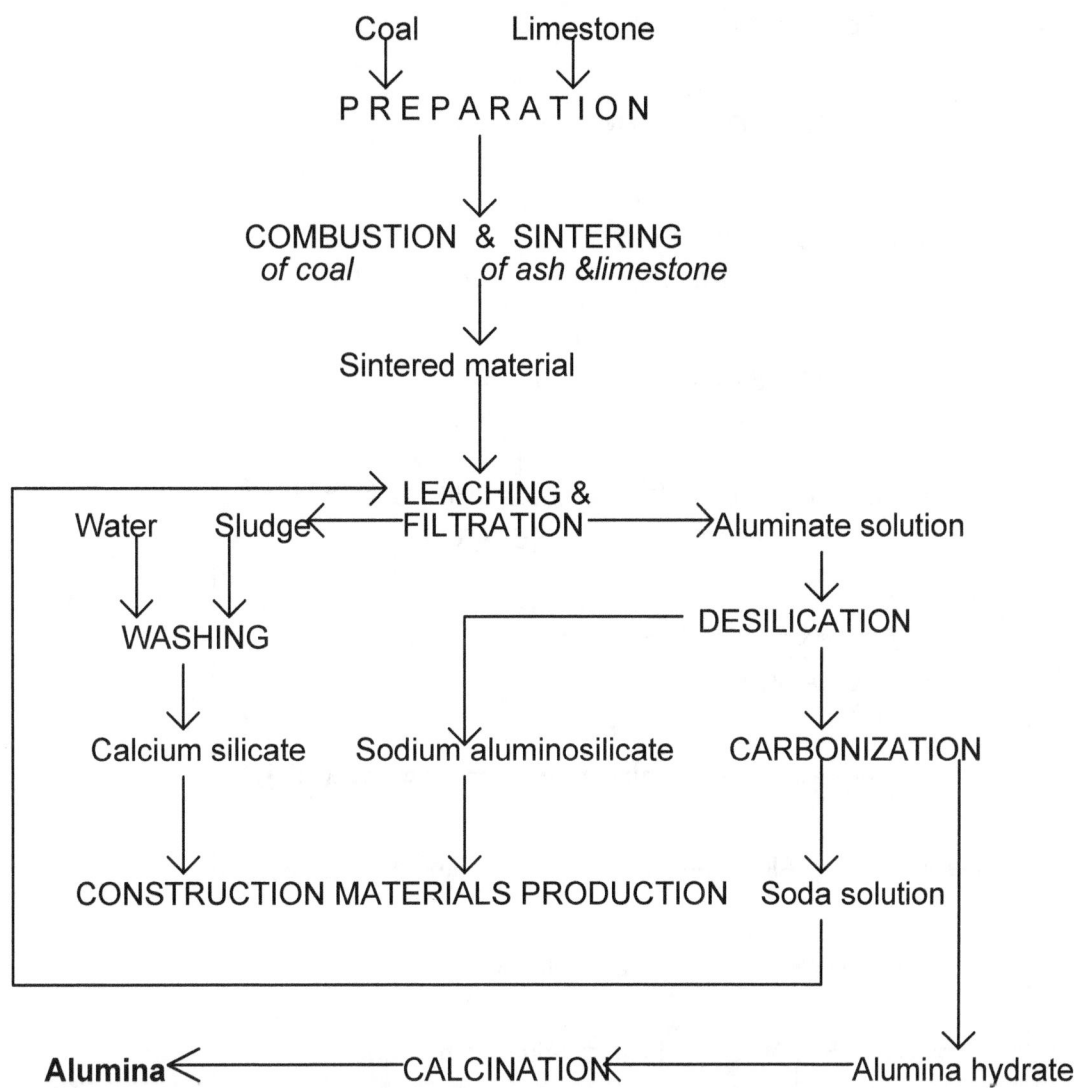

Figure 2.1.3. Flow sheet for the coal combustion and ash sintering integrated process with sintered material alkali leaching

A modification using nitric acid is shown in Figure 2.1.4.

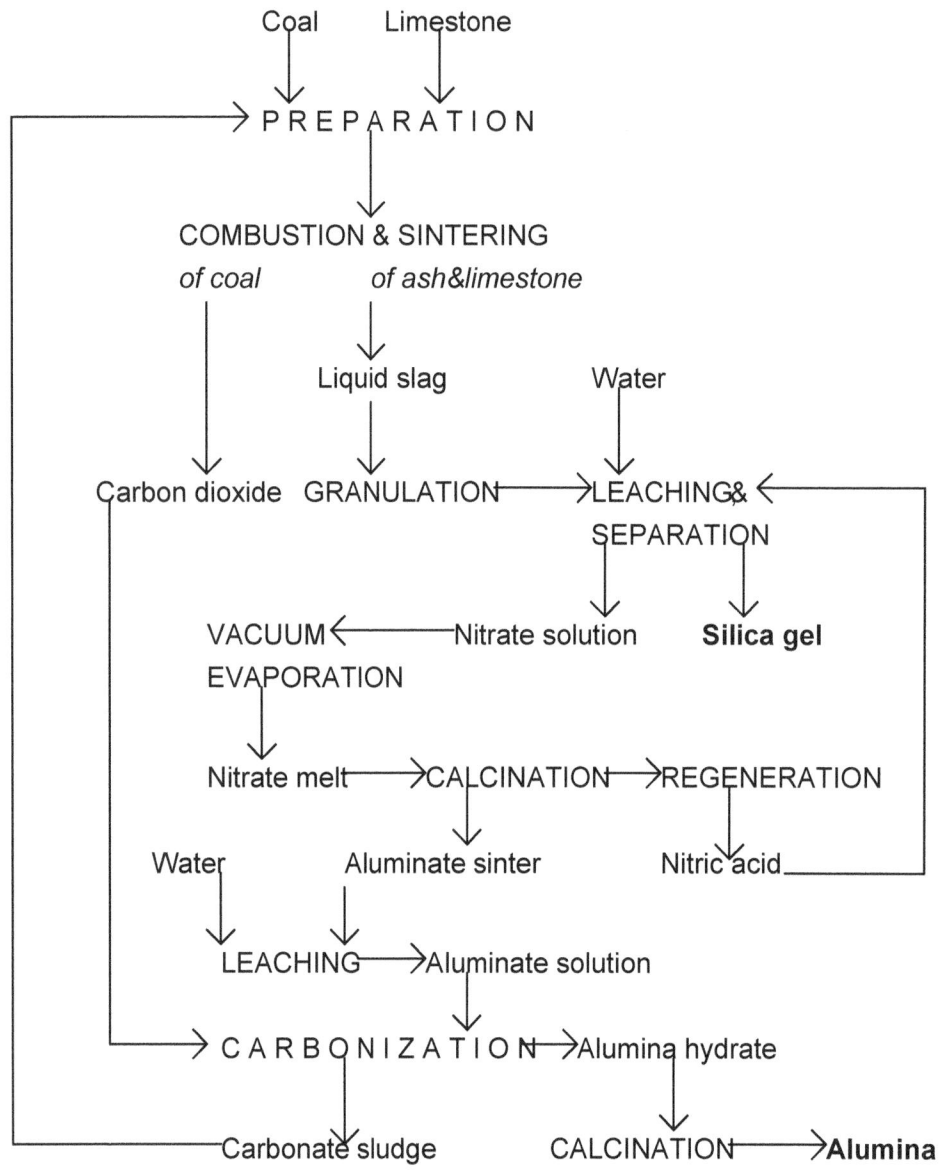

Figure 2.1.4. Flow sheet for the integrated process of coal combustion and ash sintering with sintered material nitric acid leaching

The slag obtained from the integrated process is leached with nitric acid, 25-30% HNO3, in an autoclave at 200-215°C for 3.5-4 hrs, recovering 75-80% of the Al_2O_3 in nitrate solution. Unfortunately, the alumina yield in the proposed integrated process is comparatively low due to the disparity of both combustion and sintering processes. On one hand, coal-burning time in the boiler fire-box is too short (2-5 seconds) to complete the sintering process and, therefore, limestone calcinations. There is not enough time for kaolinite conversion into metakaolinite $Al_2O_3.2SiO_2$, which is easily decomposed by the acids forming aluminum salt and silica gel. On the other hand, the high coal combustion temperature (1200°C and more) forms unleachable mullite. Combustion in a fluidized bed is favorable to desulphurization and integrated processes because of the lower temperature (700-900°C) and intensive materials stirring. An application was developed to study above technologies in a limestone fluidized bed [23]. A low-grade coal was combusted in a circulating fluidized bed to provide power for an alumina manufacturing plant at Luenen, Germany [24]. The low combustion temperature,850°C, led to the formation of minerals whose composition and crystalline structure were different from those of fly ash. The ash obtained by the fluidized bed combustion had acquired puzzuolanic properties and was used as hydraulic building material after suitable activation.

The Nucla CFB (circulating fluidized bed) Clean Coal Technology.
Demonstration Project of this method (Figure 2.1.5) is located in the Nucla Thermal Power Station of Montrose County (CO). The project aim is a demonstration of an atmospheric CFB to verify expectations of the technology's economic, environmental, and technical performance in a repowering application at an utility site to accomplish greater than 90% SO_2 removal, to reduce NO_x emissions by 60%, and to achieve an efficiency of 34% in a repowering mode. In the combustion chamber, a stream of air fluidizes and entrains a bed of coal, coal ash, and limestone. Relatively low combustion temperature limit NO_x formation. Calcium in the limestone combines with SO_2 gases, and solids exit the combustion chamber and flow into a hot cyclone. The solids separated from the gases are recycled for combustor temperature control. Continuous circulation of the coal and limestone improves mixing and extends the contact time of solids and gases, thus promoting high utilization of the coal and high sulfur capture efficiency. The flue gases passes through a baghouse where the particulate matter is removed. The steam generated in the CFB is used to generate power.

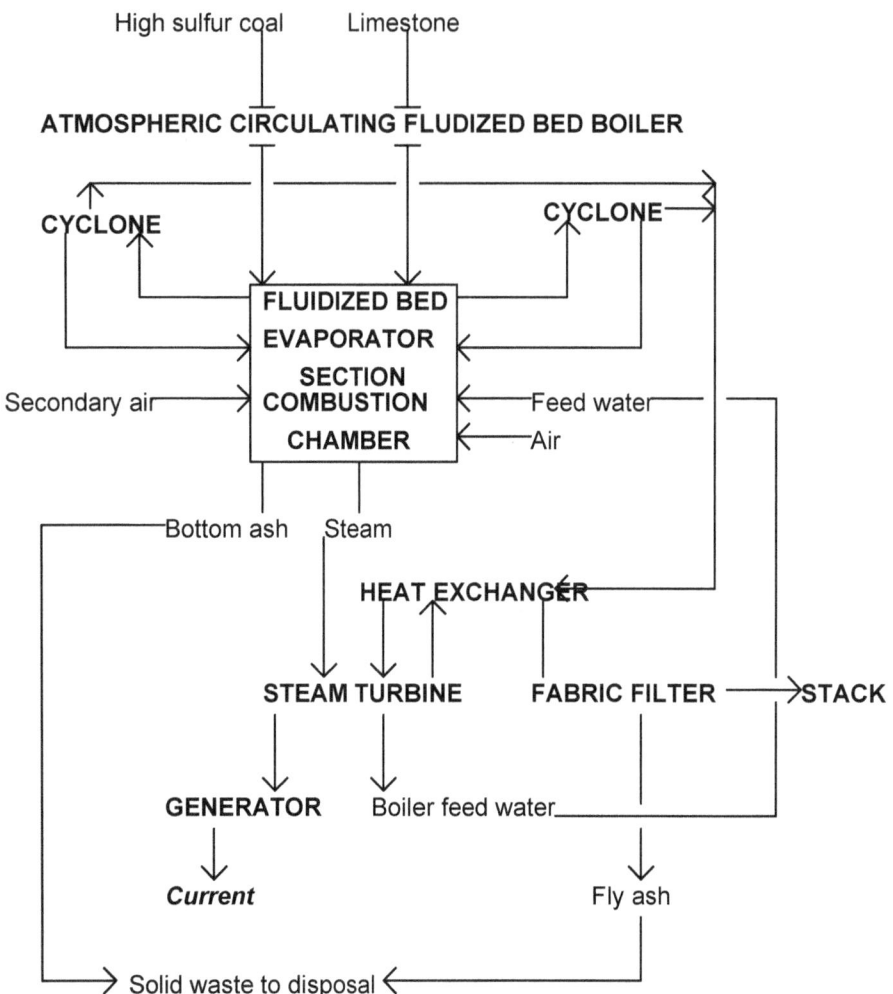

Figure 2.1.5. Flow sheet for the Nucla CFB Demonstration Project

Acidic Leaching Coal Waste. Any mineral acid can be used for digestion of the aluminum-containing raw material; however, only sulfuric, hydrochloric, nitric, or hydrofluoric acids have practical significance. A flow sheet for processing coal ash by sulfuric acid into aluminum sulfate is shown in Figure 2.1.6.

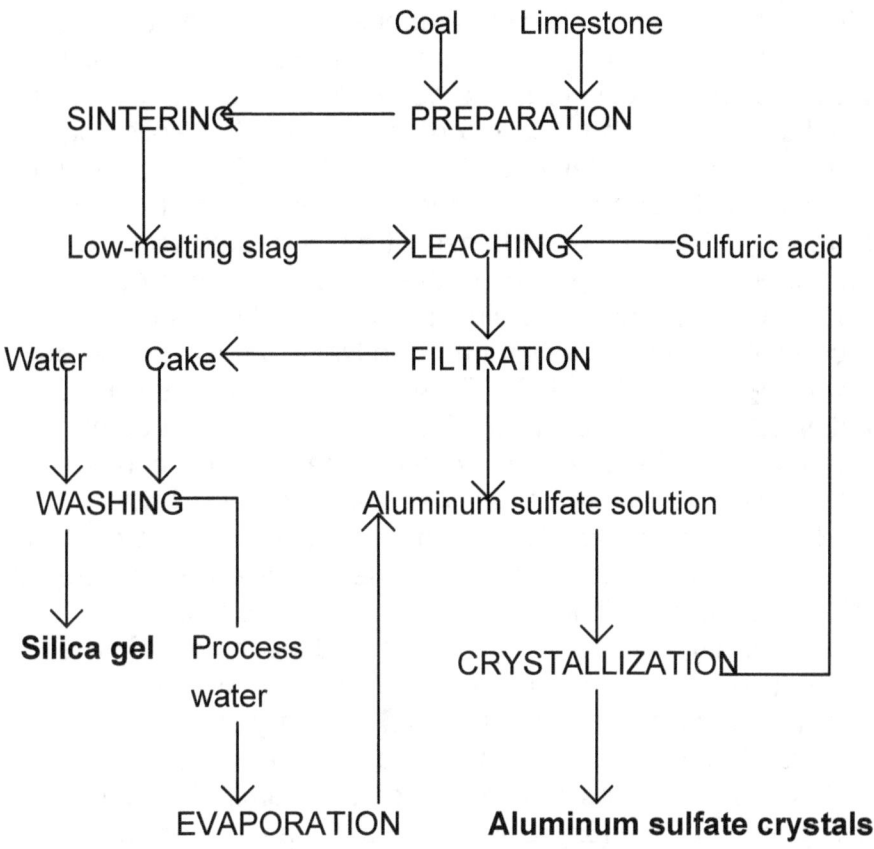

Figure 2.1.6. Flow sheet for processing coal ash by employment of sulfuric acid

This method was tested by sintering a coal ash with limestone at 1300°C for 2 hrs followed by leaching the sintered material with 2 N sulfuric acid to give an 82% recovery of alumina in liquor [25]. Fly ash was initially subjected to magnetic separation to remove the magnetite portion. The nonmagnetic fraction was treated with a calcium-containing agent so that the $CaO:SiO_2$ ratio became 1.4-2.0:1, and water was added to form 1.25 cm pellets. After the pellets were calcined at 1100°C, they were treated with H_2SO_4 to form an aqueous solution of $Al_2(SO_4)_3$ and a suspension of $CaSO_4$ and $CaSiO_3$. The aluminum sulfate and subsequently alumina were recovered by conventional procedures. Solvent extraction was used to purify the aqueous $Al_2(SO_4)_3$ [26,27]. The calcined material was cooled to 350°C, then introduced into a reactor and leached with H_2SO_4 for 30 min at 90°C. The highest Al extraction was 99.4%, which was obtained by leaching with 2.5 N H_2SO_4 under hydrothermal shock force. The aluminum sulfate was crystallized from solution and calcined to yield alumina which was approximately 99.9% pure. Alumina was also recovered from the coal combustion fly ash by sintering a 1:1:1 mixture of the ash + $CaSO_4$ + $CaCO_3$ at 1000-1500°C to convert the ash components to a crystalline form of H_2SO_4-leachable aluminum together with other incidentally recoverable metals. The depleted liquor was treated with

Primene JM-T to extract Fe, Ti, U, and Th. The aqueous raffinate was used for recovering alumina [28]. The coal ash can be ground with water to form a slurry from which a magnetic fraction is recovered. The nonmagnetic fraction could be leached with boiling sulfuric acid for the recovery of metals, residual solids rich in CaO, and silicates filtered from the metal-loaded acidic solution [29]. That solution can be treated with sodium hydroxide for a controlled precipitation of Ti and Fe hydroxides with recovery by filtration. The obtained filtrate is treated with CO_2 or powdered $Al(OH)_3$ to promote the precipitation of the aluminum hydroxide for recovery and subsequent calcinations to the alumina. This product can be produced from the power plant fly ash by sintering with NaCl and Na_2CO_3 at 500-900°C and leaching with H_2SO_4 or HNO_3 [30]. Three parts of sinter medium (2 parts NaCl and 1 part Na_2CO_3) were mixed with 1 part fly ash and heated at 400°C for 1 hr and then at 850°C for 2 hrs. The sintered material was cooled, leached with water, and filtered, and the residue was leached at 85°C with 1 N HNO_3 to recover 98.7% of the aluminum. The leached solution was extracted with a 50% solution of bis(2-ethylhexyl)phosphoric acid in diethylbenzene to recover all the Fe and Ti and 95% of the U and Th in the leach. Sintering coal ash with limestone was optimized for alumina recovery by nitric acid leaching [31]. The solubility of Al_2O_3 was related to the thermal decomposition of mullite in the presence of limestone in a crucible furnace. Maximum recovery of aluminum was attained after sintering at 1220-1250°C, for 50 min and more, and with a $CaO:Al_2O_3$ ratio of 1.45-1.60:1. Besides alumina, a sodium aluminate can be produced by processing the coal ash [32]. One part of ash was treated with 2 parts of water and 1 part of 98% HNO_3 at 65°C, and the product was filtered. $CaSO_4$ was precipitated by treating 2.5 parts of filtrate with 0.2 parts of 0.5% H_2SO_4. The residue was treated with NH_4OH to precipitate aluminum and iron hydroxides which were then separated by filtration. The solids were treated with NaOH and filtered to give a solution of the sodium aluminate which was then evaporated to crystallize the solid product with 97.12% purity (Figure 2.1.7).

It is also possible to prepare ammonium alum $AlNH_4(SO_4)_2.12H_2O$ from power plant ash by treatment with HNO_3 and H_2SO_4, neutralization of the product with NH_3, cooling, and filtration to recover the product [33,34]. There is a process for recovery of gallium and germanium from coal ash [35]. The ash is palletized and heated in an oxidizing atmosphere at 900°C and higher but below the fusion point of the pellets to remove volatile trace elements. The pellets are treated at the above temperature in a reducing atmosphere to transfer the Ga and Ge oxides to suboxides which sublime. Then both above suboxides are recovered from the gas. The less common metals can be recovered from the calcium-rich ash after combustion of brown coal by pressure leaching with heating by hot flue gases [36].

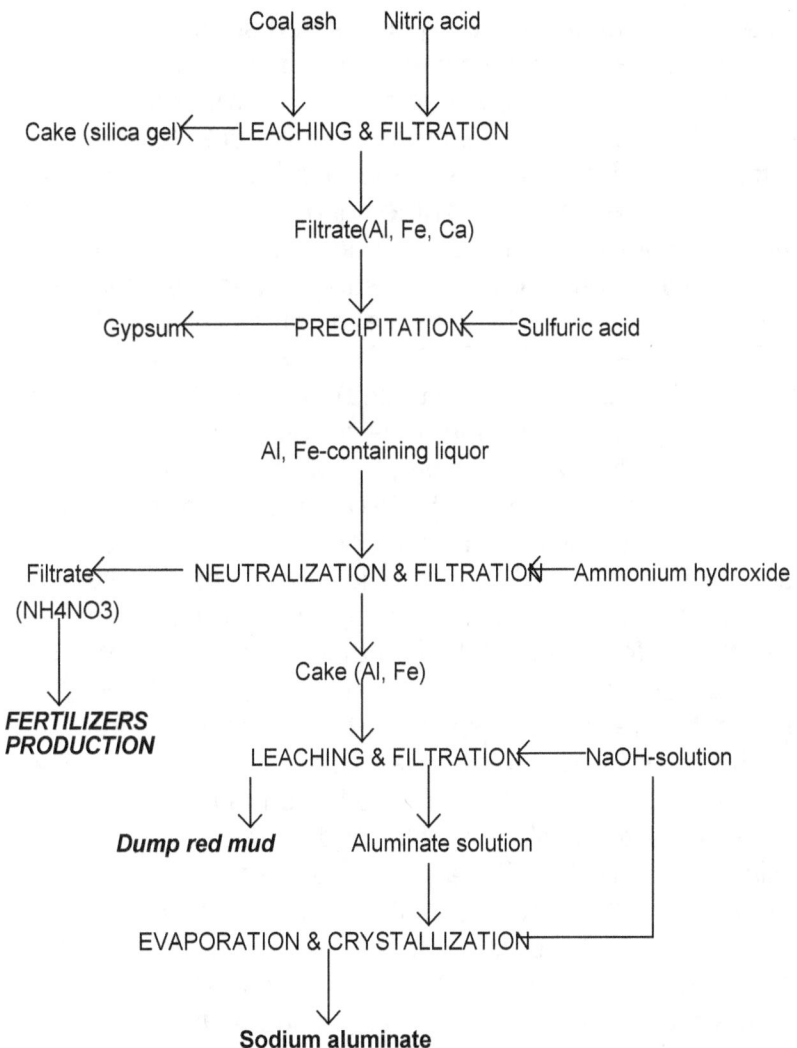

Figure 2.1.7. Flow sheet for sodium aluminate production

The leaching solution is concentrated, and the metals are separated by conventional ion exchange and electrolysis. Fine tailings from collieries and coal ash were evaluated as feed for some metals recovery by hydrochloric acid leaching. The rates of extraction of aluminum, iron, and potassium were measured under standard calcinations and leaching conditions [37]. Results were expressed in terms of the chemical, mineralogical, and physical properties as well as the thermal histories of the feed materials. It was concluded that coal ash from combustion at more than 850°C are not suitable for alumina recovery. Controlled calcinations at 650-850°C was necessary for conversion of the kaolinite into the more reactive metakaolinite. The solubility of Al_2O_3 decreased sharply at temperatures higher than 900°C; therefore, calcinations in a fluidized bed combustor is the only industrial scale method suitable [38]. The rate of extraction of Al and Fe from coal ash and colliery tailings increased with increasing temperature and acid concentration [39].The rate of the aluminum extraction was higher using HCl than that with HNO_3 under the same conditions. The removal of the Fe from acidic HCl leach solutions is achieved by adding a tertiary amine, Alamine 336 [40]. The optimum temperature for Ekibastuz coal ash leaching with hydrochloric acid was determined [41,42]. The tested ash contained the following (in wt %): Al_2O_3 28.43, SiO_2 58.0, Fe_2O_3 4.74, CaO 3.37, TiO_2 1.3, MgO 0.4, and Na_2O + K_2O 0.72. Recovery of Al (calculated as alumina) increased from 57.71% at 150°C to 66.6% at 260°C. The optimum temperature for autoclave leaching was 180°C at HCl concentrations of 15-20% and leaching times of 2-2.5 hrs, and at 100-115% of stoichiometric acid consumption [43]. The conditions for recovering some constituents from the coal waste were examined [44]. Fly ash was leached at 90-135°C with a solution at pH 11-14 and filtered. The solids recovered were treated with HCl or HF to form a solution containing aluminum and iron chlorides or fluorides. The Fe was removed from the solution by electrodeposition at pH 1-3, a temperature of 70-90°C, a potential of 1.5-3 V, and the current density of 1.2-1.5 A/in2. Silica was filtered from the electrolyzed solution. The pH of the Si-free solution was adjusted to 6.5-7.5 at 30-90°C to recover aluminum hydroxide. The recovery was SiO_2 100%, Fe 65%, and Al_2O_3 89%. The Si-containing residue from acid leaching can be mixed with graphite powder and heated in argon at 1600-1700°C to obtain silicon carbide [45]. The ash collected from coal-fired furnaces was leached with HCl and the leached solution was extracted with a tertiary amine at ambient temperature to remove Fe contaminants. The purified solution was injected with gaseous HCl to precipitate a crystalline $AlCl_3.6H_2O$. The residue was decomposed to obtain Al_2O_3 [46]. The direct acid leaching (DAL) process deserves attention [47]. The flow chart is shown in Figure 2.1.8.

According to this method, ash containing (in wt %) SiO_2 52.31, Al_2O_3 27.18, Fe_2O_3 10.81, MgO 1.04, Na_2O 0.23, K_2O 0.79, TiO_2 1.46, MnO 0.03, P_2O_5 0.25, BaO 0.12, SrO 0.07, SO_3 0.51, C 2.61, and LOI 3.06 is treated with 6 M HCl at 110°C under reflux. Ash is classified into two portions. The first portion (60%) contains the fine fraction (95%, - 35µ), and the second portion (39.5%) contains the fraction with larger particles (2%, - 20µ). The fine fraction portion was subjected to magnetic separation, and then the nonmagnetic material was treated with HCl. As a result, 39% of Al and 79% of Fe were extracted. The total quantities of recovered aluminum and iron in terms of their losses with the nonmagnetic fraction were equal to about 24% and 30%, respectively. However, the authors of the DAL process did not aim at high extraction percentages of individual metals. Their purpose was to convert ash into various valuable products. The remainder of the material can be used as a good plastic filler (DAL-filler). Iron is removed from an aluminum chloride solution by passing it through anion-exchange columns and then treating it with gaseous HCl. In such manner, aluminum chloride hydrate $AlCl_3.6H_2O$ is produced. This product is calcined into alumina 12.5%, ferric oxide 8%, gypsum 3.5%, alkali

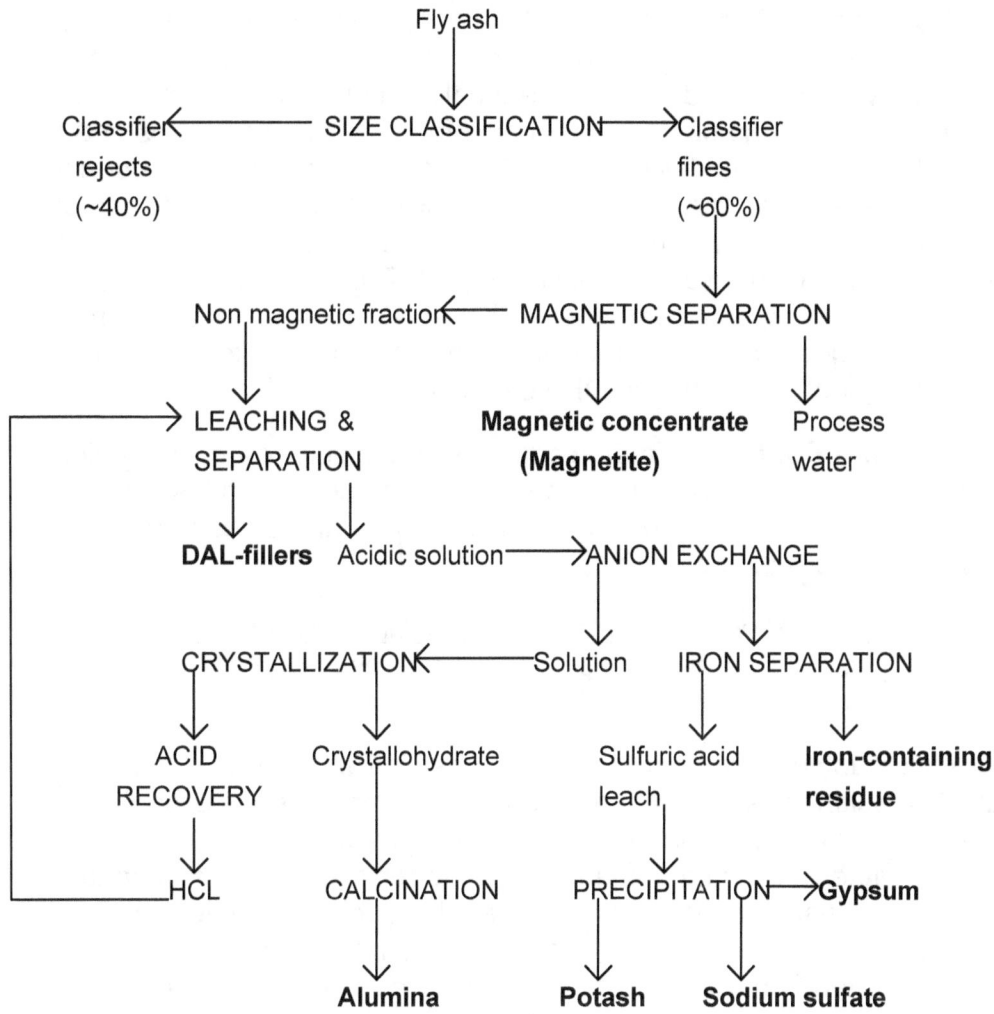

Figure 2.1.8. Simplified DAL process diagram

sulfates 6%, spent ash 68%, and others 2%. The mullite and amorphous aluminosilicate contained in coal ash which are relatively insoluble in HCl, are easily dissolved in a HCl-HF mixture. That liquor was used for direct acid leaching of some samples of coal ash [48]. The dissolved Si was vaporized as halide gases such as SiF_4 and $SiCl_4$. These gases were recovered in absorption columns. All coal ash tested showed more than 90% Si vaporization. Also, fly ash reacts easily with hot aqueous $HSiF_5$ and HF to form SiF_4 vapor that can be separated by distillation at 90-100°C from the dissolved fluorides and fluorosilicates rich in Al and Fe [49].

The recoverd SiF_4 was hydrolyzed to form SiO_2 thereby recycling the HF. The salt-rich solution was passed through an electrolytic cell for Fe recovery. The resulting solution was processed to precipitate calcium and magnesium salts and to purify aluminum salts. This process is suitable for the leaching of ores and the recovery of metals from mining tailings. Bright white calcium silicate compounds, useful as a filler material, can be produced by the acidic treatment of fly ash or scrubber sludge, filtration of the slurry obtained, and subsequent mixing of the filtered cake and ammonium hydroxide solution [50]. The remaining solid residue is the calcium silicate. Magnetic components, especially magnetite, are removed before extraction. Crude $Al(OH)_3$ is precipitated from the acid liquor and purified by leaching with aqueous NH_3 and NaOH solutions.

Alkaline Processing Coal Ash. Technologies using alkali for processing high-silicon aluminum-containing raw material and waste, including coal ash, are classified as follows: 1)sintering processes, 2)hydrothermal methods, and 3)methods with silica pre-extraction [51].

Sintering Method for the Processing Coal Ash into Aluminum and Silicon-Containing Products. A flow sheet of the sintering process is shown in Figure 2.1.9.

The coal ash is ground with limestone and soda ash in definite proportions. The sintering of the resulting blend is based on soda and limestone decarbonization; silica and calcium oxide binding in larnite (β-dicalcium silicate) β-Ca_2SiO_4 which is insoluble in aluminate liquor; and sodium aluminate formation. The aluminate solution obtained by leaching the sintered material is subjected to desilication [52] and is further decomposed by carbonization. Aluminum hydroxide, after being precipitated, separated from the soda solution and washed, is processed into alumina. The slurry, containing β-dicalcium silicate, is utilized to obtain cement. The alumina yield from coal waste obtained by the limestone-soda sintering process was about 88%. The optimal conditions included sintering at 1250-1300°C for 20-30 min, using a Na_2O:Al_2O_3 ratio of 1.2-1.3:1 and a CaO:SiO_2 ratio of 1.8-2:1, and leaching at 60-70°C for 20-30 min with 5% soda solution [53,54]. Leaching the sintered material with water gives a lower SiO_2 content than leaching with NaOH and Na_2CO_3 [55,56]. The obtained sodium aluminate solution is processed with $Ca(OH)_2$ suspensions at atmosphere pressure to desilicate, reducing the SiO_2:Al_2O_3 ratio to less than 1:1000. Fly ash was sintered with limestone and soda to form Ca_2SiO_4, and about 90% of the Al_2O_3 was recovered as $NaAlO_2$ [57,58]. No apparent difference in reactivity was found between the Ca contained in the fly ash and that in the added limestone [59].

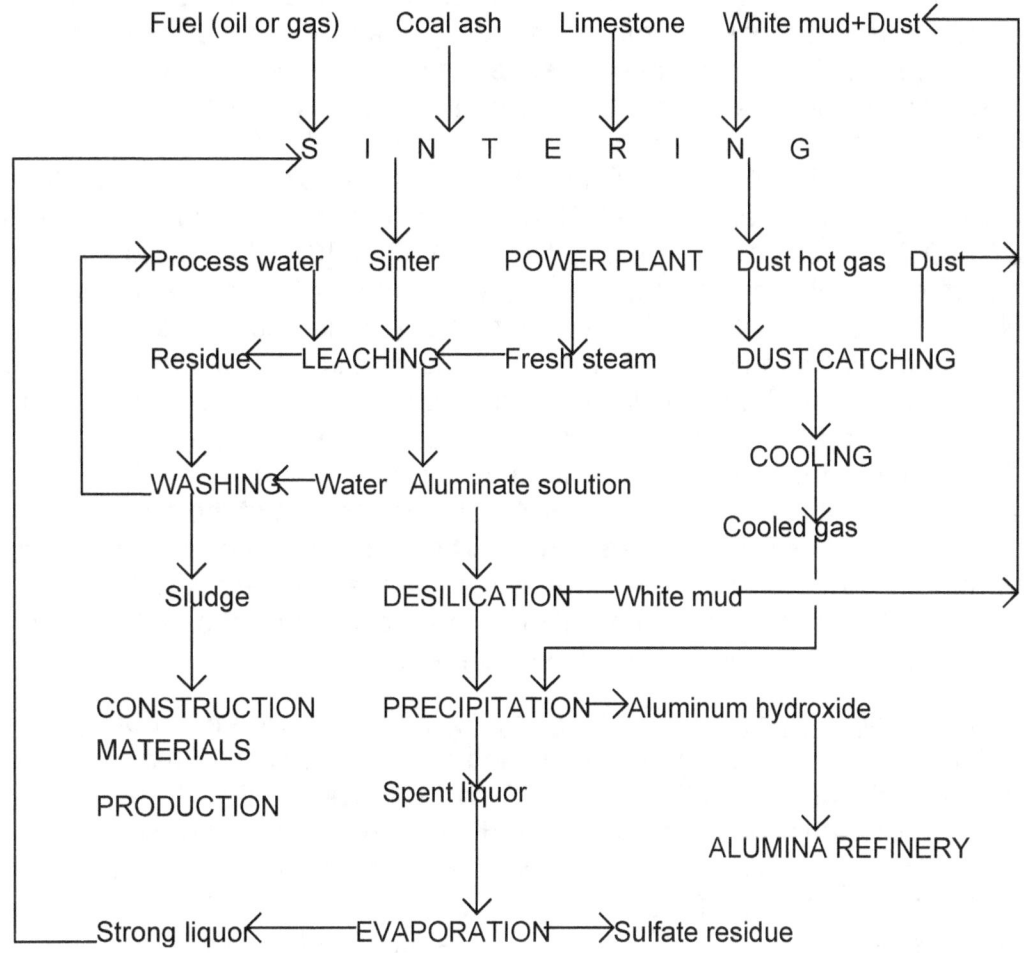

Figure 2.1.9. Flow sheet for the sintering process of coal ash utilization

The Al_2O_3 yield was increased, and the sintering temperature was decreased by adding 2-10 wt % of carbon and sulfur at S:C ratio of 0.5-1.5:1 and heating at 1100-1300°C. In the absence of C and S, the sintering temperature must be 1350-1300°C [60]. More than 90% of the Al_2O_3 was recovered from the sintered material by soda solution leaching. Such high levels of alumina yield were achieved for both bituminous and subbituminous coal ashes. The obtained calcium silicate was used as a feed for cement production [61]. Fly ash may be sintered at 1140°C for 60 min with $CaCO_3$ and K_2CO_3 additives (the $CaO:SiO_2$ and $K_2O:Al_2O_3$ molar ratios are 2:1 and 1:1, respectively). All Al_2O_3 will be converted to H_2O-soluble aluminate [62]. High-quality Al_2O_3 can be produced from schist and fly ash using a process in which the melt self-pulverizes on cooling due to the volume change of the dicalcium silicate phase transition. The powder should be leached and SiO_2 removed. Aluminum hydroxide is then precipitated and calcined to give alumina [63]. The sludge filtered from the slurry obtained from the coal waste and limestone mixture sintering is a raw material for cement production [64]. In the limestone-coal ash sintering process, the Ca_2SiO_4 - and $Ca_{12}Al_{14}O_{33}$ -bearing sinter is cooled and leached with a Na_2CO_3 solution at 50-70°C. The solids are removed from the leach by settling and/or filtration. The resulting liquor is treated with slaked lime, and SiO_2 is precipitated by heating to 90-120°C with agitation. After removal of the silicate-containing solids, the liquor is cooled to70-90°C and carbonized by CO_2 to form $Al(OH)_3$ [65]. If a fluidized bed is used for limestone-coal ash sintering, then agglomeration of bed material and ash can occasionally cause problems [66]. Addition of limestone decreases the sintering ability of coal when the Ca:Si ratio of 3:1 is exceeded. The decrease was even greater when an Al-Si-based clay mineral was used. Besides cement, silica brick may also be produced from a byproduct in the limestone-coal ash sintering process [67]. The product is obtained by the hydro chemical treatment of the Ca_2SiO4 -containing sludge with an alkali solution, preparation of the molding sand, and subsequent hydrothermal treatment. The strength of the raw material is increased and the bulk density of the product is decreased by the hydrothermal treatment of the sludge with an alkali solution with a Na_2O:SiO_2 ratio of 0.5:2.5 at 50-105°C for 1-24 hrs with subsequent washing and drying.

Hydrothermal Alkaline Processing of Coal Ash. Hydrothermal treatment of high-silica aluminum-containing raw material is based on an interaction of the aluminoailicate with caustic solution and lime in an autoclave at 250-300°C. Silica from the aluminosilicate is bound to CaO in sodium monocalcium hydro- silicate $NaCaHSiO_4$, resulting in half the limestone consumption compared to the sintering process [68-71]. A flow sheet of the hydrothermal process (Figure 2.1.10) includes the following stages of alumina recovery: (1) preparation of the initial slurry by mixing coal ash, lime, sodium hydroaluminosilicate (SHAS) precipitated by desilication of an aluminate solution, and the depleted caustic liquor of the sodium hydroaluminate crystallization, (2) digestion of the prepared slurry and subsequent separation by filtration into the alkali-aluminate solution and the $NaCaHSiO_4$-containing sludge, (3) recovery of alkali from the filtered sludge by water washing, (4) desilication of the alkali-aluminate solution and precipitation of the formed SHAS, (5) evaporation of the aluminate-containing solution followed by cooling to crystallize sodium hydroaluminate (SHA), and (6) dilution of the SHA, and the precipitation of aluminum hydroxide and its calcinations into alumina. The impact of various conditions on alumina recovery from coal ash was examined at temperatures between 100 and 350°C, alkali concentrations of 100-500 g Na_2O/L, molar ratios of Na_2O:Al_2O_3 of 7-20:1 and CaO:SiO_2 of 0.7-1.3:1, and residence times of 1-60 min. As a result, the following optimal parameters were determined: 270-280°C, 400-500 g Na_2O/L, molar ratios of Na_2O:Al_2O_3 of 12:1 and CaO:SiO_2 of 1.1:1, and 15 min of duration [72].

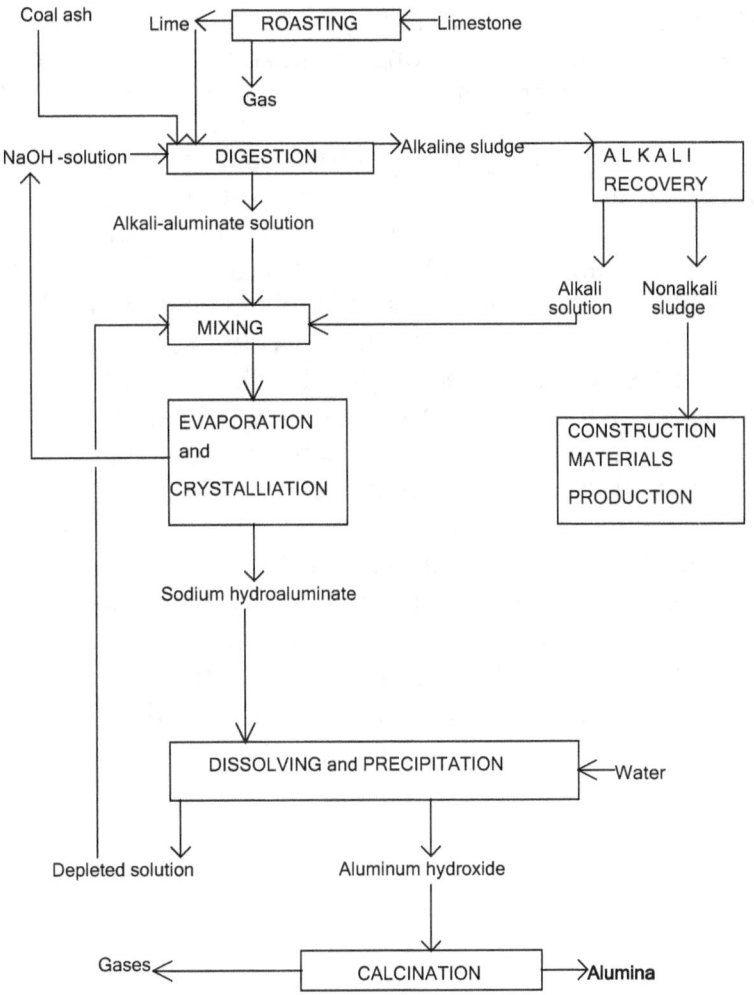

Figure 2.1.10. Flow sheet of the hydrothermal process for recovery of alumina and byproducts from coal ash

The energy and fuel consumption for alumina recovery from coal ash and other no bauxite raw materials and aluminum-containing wastes can be decreased by hydrothermal alkali leaching in the presence of lime in a tubular autoclave [73]. Leaching using a reclaimed alkali-aluminate solution of high $Na_2O:Al_2O_3$ ratio (20-50:1) gave an aluminate-alkali solution with a $Na_2O:Al_2O_3$ ratio of 10-12:1 and Al-free residue containing sodium-calcium hydrosilicate (SCHS) $NaCaHSiO_4$, from which NaOH can be readily regenerated. Conditions of the sodium hydroaluminate (SHA) crystallization from concentrated solutions obtained from Ekibastuz coal ash leaching were as follows: temperature 45°C, hold-up time 8 hrs, Na_2O concentration in the evaporated solution 500-700 g/L, $Na_2O:Al_2O_3$ ratio in an initial solution 6-7:1, and in depleted liquor 30:1. The alumina precipitation into SHA under these conditions was 84% [74]. Zeolite A may be prepared from alkali metal silicate solutions by leaching silicate components of coal ash with

NaOH liquor and from the aluminate solution obtained by hydrothermal digestion of ash in the presence of lime. The combined solutions are mixed and heated to precipitate an aluminosilicate gel which is crystallized to give the final product [75].

Alkaline Pre-Extraction of Silica from Coal Ash. The method assumes that a significant part of the ash silica is in glassy phases and cristobalite which are alkali soluble. The aluminum of the coal ash is in the unleached mullite and partly in clay mineral like kaolinite [76]. Therefore, the solid residue after alkali treatment of ash is enriched with Al_2O_3 [77]. At the outset, coal ash is mixed with an alkali solution, and this suspension is heated and stirred at 80-105°C to extract soluble silica (Figure 2.1.11). The resulting slurry is filtered to give an alkali-silicate solution and an alumina-containing cake [78,79]. The separated solution is desilicated by lime addition to form calcium hydrometasilicate and reusable alkaline liquor. The alkali-silicate solution can be processed into a wide assortment of valuable chemicals, such as sodium and calcium hydrometasilicates, soda-silica mixtures, both amorphous and crystalline silica, zeolites, glasses, ceramics, and other inorganic and organic silicon-containing materials. Ash from Ekibastuz coal combustion was treated with alkali solution in an autoclave. In several cases the SiO_2 content was decreased from 59 to 30% with a simultaneous increase in the Al_2O_3 content from 26 to 39% [80-82].

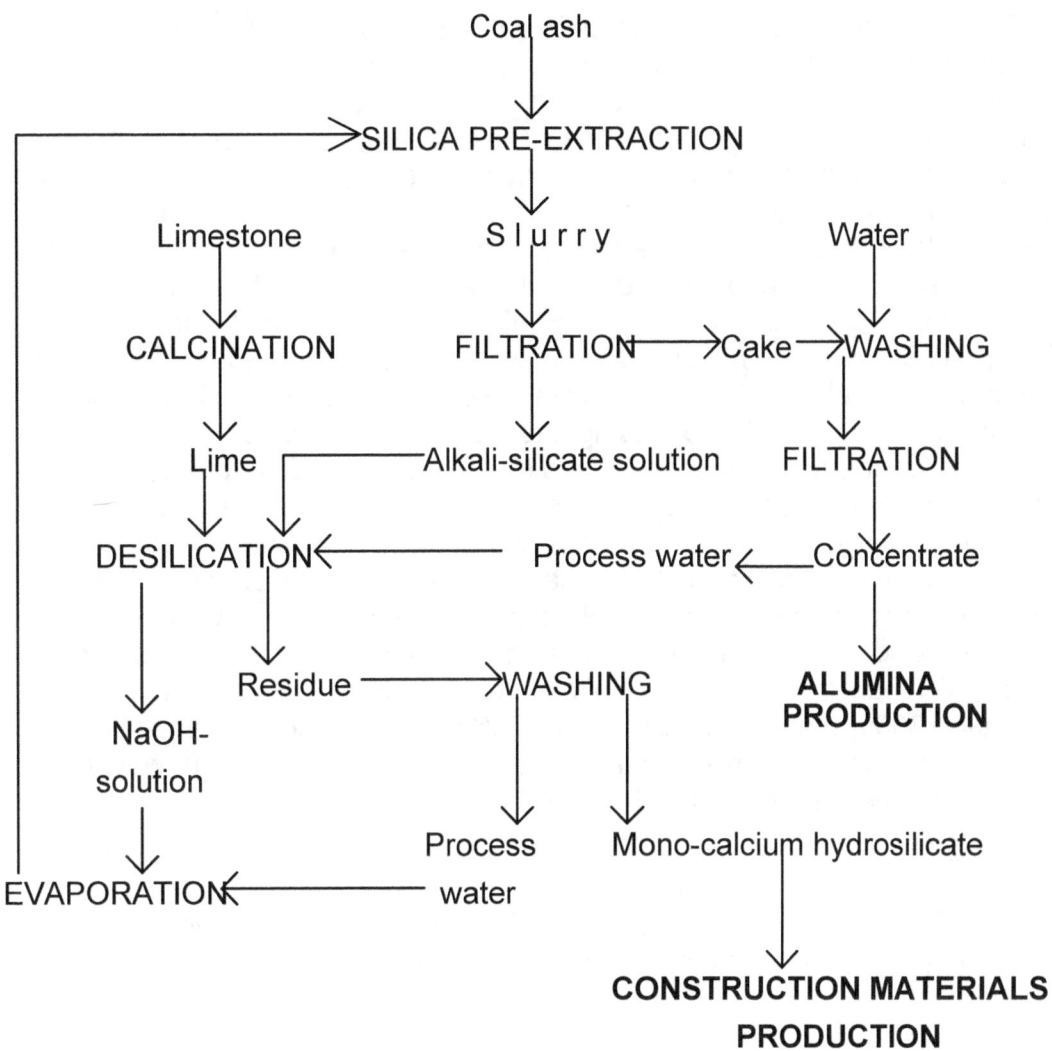

Figure 2.1.11. Flow sheet for the process of alumina production from coal ash

Pilot Plant Tests and Prospective Technology for Processing Coal Ash and other Nonbauxite Raw Materials into Alumina and Byproducts. Some of reviewed processes were carried through the pilot plant test stage. Coal ash, schist, clay, and other nonalkali aluminosilicate raw materials were used as an alumina source. In 1977 a pilot plant in Estak, France, that turns out 15,000-20,000 tons of alumina yearly was put into operation to test the so-called **H+ process** [83]. According to this method, the raw material (oil schist) was ground and roasted at 700°C and then leached by 50% H_2SO_4 solution at 130°C and atmospheric pressure. The obtained sulfate aluminic liquor was treated by hydrochloric acid to form aluminum hydrochloride that was subjected to thermal decomposition into alumina. It was found that investments and operating expenditures, which were 15% greater than the Bayer process, were compensated for by reductions in bauxite import. The result is that the raw material share of the cost of produced alumina is 45% for the Bayer process and 40% by the *H+ method*. Another pilot plant was designed in Boulder City, Colorado, U.S. in 1973. Clay and anorthosite were tested using of acidic extraction and limestone-sintering processes [84]. A comparison of estimated expenditures for the six tested processes was conducted [85]. It was concluded that *HCl-extraction* process is the most promising and viable for implementation. In comparison with this method, the other investigated processes were characterized by the following additional investments and the operational costs (per 1 ton of alumina): *HNO₃ -extraction* $319.4 and $74.5, *H₂SO₄-extraction* $565.4 and $39.5 and sintering process $407.5and $54.5. The comparison uses early 1980 dollars. It was asserted [86] that a hydrothermal alkaline method for coal ash processing into alumina has been under development for about 10 years at Alcoa's small pilot plant [87]. It was reported that the cost of alumina produced by the above method is nearly equal to the Bayer process involving bauxite. Also, the hydrothermal alkaline method was pilot plant tested at the specially erected tubular autoclave unit in Czechoslovakia using some low-grade aluminum-containing raw materials including coal ash [88]. All of the feedstocks were digested to ensure a fair technical and economical evaluation. The parameters were as follows: temperature 300°C, Na_2O concentration 300 g/L and molar ratio of $CaO:SiO_2$ 1.1:1. The residence time of the slurry running through an autoclave system from pump to drossel was 10 min. The results obtained are shown in Table 2.1.2.

Table 2.1.2. Results of pilot plant test of the hydrothermal alkaline-lime method for processing various nonbaxite aluminum- and silicon-containing raw materials

	clay 1	clay 2	clay 3	kaolin 1	kaolin 2	shale clay	shale	coal ash
Composition (wt%)								
Al_2O_3	34.0	26.6	21.4	32.0	34.0	31.6	18.8	30.2
SiO_2	43.0	32.2	57.1	52.8	48.0	49.9	52.8	50.6
Fe_2O_3	3.0	16.3	5.1	0.9	1.3	2.8	8.8	8.4
CaO	0.5	0.3	0.6	0.3	0.5	0.4	1.0	2.6
Na_2O	0.03	-	0.11	0.02	0.06	0.04	1.0	0.3
K_2O	0.09	-	2.24	0.73	1.74	1.02	-	0.4
LOI	15.0	20.0	9.0	12.0	12.0	11.7	9.0	4.0
$H_2O_{(l)}$	15.0	15.0	20.0	10.0	10.0	10.0	-	-
Main constituents								
kaolinite	x	x	x	x	x	x	x	x
quartz	x	x	x	x	x	x	x	x
mica	x		x	x	x	x	x	
montmorillonite	x		x	x		x		
hematite	x		x			x	x	
siderite		x	x					
feldspar			x		x		x	

40

Table 2.1.2. (completion)

	clay 1	clay 2	clay 3	kaolin 1	kaolin 2	shale clay	shale	coal ash
Yield (%)								
Al_2O_3	85	60	50	85	85	80	70	80
Na_2O	75	85	85	75	75	85	70	85
Material consumption (ton per 1 ton Al_2O_3)								
raw material	4.38	8.44	12.69	4.37	4.09	4.75	8.25	4.47
lime	1.98	2.60	6.52	2.47	2.14	2.55	4.96	2.69
caustic	0.28	0.34	0.55	0.35	0.31	0.24	0.66	0.28
water	8.63	10.94	17.6	9.97	9.59	10.88	15.40	12.34
Energy consumption (per 1 ton Al_2O_3)								
steam (GJ)	12.61	20.92	38.18	14.81	13.26	15.53	27.31	16.27
fuel (GJ)	26.61	32.29	36.68	26.72	26.58	27.51	30.08	27.30
power (kWh)	1400	1800	2300	1400	1400	1500	1800	1500
Output of dumped sludge (ton per 1 ton Al_2O_3)								
	4.90	8.77	17.52	6.03	5.42	6.73	13.08	7.56
Investments (%)								
	100	130	167	103	100	107	130	107

High Al_2O_3 yields (80-85%) were reached by digestion of raw materials with Al_2O_3:SiO_2 ratio more than 0.6:1, apart from the clay (sample 2) high in iron oxide (16.3%) as siderite. Fifty four tons of Ekibastuz coal ash were subjected to alkaline pre-extraction of silica at the Russia's pilot plant [89-91]. The obtained concentrate contained 50.2% of Al_2O_3 and 29.9% of SiO_2 against 27.1% of alumina and 60.2% of silica in the starting ash. The SiO_2 yield into alkali-silicate solution came to 72-74%, which is 87-89% of the theoretical yield. The improved raw material was sintered with soda and limestone. 86% of Al_2O_3 and 90% of Na_2O were recovered from the sinter to aluminate solution. The previously filtered alkali-silicate solution was desilicated by the addition of lime in amounts of 250-300 g CaO/L at temperature of 95-100°C for 6-8 hrs. The calcium hydrosilicate as plombierite was filtered from the reclaimed alkali solution. The solids were successfully tested for manufacturing silicate brick and portland cement [92]. Technical and economical estimates of the conducted pilot plant test data [93] suggest that the silica pre-extraction causes a decrease in investments (6.7%) and alumina price (14.5%) in comparison with convention limestone-soda sintering method.

Discussion Concerning the Reviewed Processes. All of the processes developed and applied to coal ash utilization may be classified in two groups. The first involves the methods which are not destined for ash constituents recovery. About 22% of the U.S. coal ash is applied directly as an admixture to plastics ("High volume use"), cement or concrete ("Medium technology uses") [3] and others. The second group includes technology intended for ash constituents recovery. This class of processes, called "High technology ash application" [3], can be broken down into two sectors. The direct recovery of metals from ash through aqueous acidic or alkaline treatment, sintering followed by sinter leaching, integrated coal combustion and ash-limestone sintering is the first sector of "High technology". The methods for silica pre-extraction from coal ash followed by the limestone-soda sintering technology make up the second sector. It is known that kaolinite, being a principle aluminum-containing mineral of coal, is converted during coal combustion through metakaolinite to mullite and cristobalite as follows [94]:

$$3Al_2Si_2O_5(OH)_4 \rightarrow 500\text{-}900°C \rightarrow 3Al_2Si_2O_7 + 6H_2O \rightarrow$$

    kaolinite   *metakaolinite vapor*

$$3Al_2Si_2O_7 \rightarrow 900\text{-}1300°C \rightarrow Al_6Si_2O_{13} + 4SiO_2$$

metakaolinite   *mullite*   *cristobalite*

The products of the above chemical transformation vary in their reactivity to alkalies and acids. Kaolinite is poorly decomposed by acid [94]. In contrast to this, kaolinite is reactive to alkali [76]. Metakaolinite formed as a result of heating kaolinite to 500-900°C is characterized by significant reactivity to both acidic and alkaline solutions. However, the 500-900°C temperature range of coal combustion is only typical of fluidized bed combustors [5]. Therefore, processing of colliery tailings by acid-leaching routes using the FBC is the only suitable commercial scale method [38].

Unfortunately, the great bulk of coal is burned at temperatures significantly higher than 1000°C. Consequently, the ash consists of mullite that is both alkali and acid resistant. Coal combustion products contain 44% bottom ash and 47% fly ash [89]. Leaching it out at high temperature (250°C or more) requires autoclave technology. It should be remembered that employment of acid for processing of aluminum-containing ash is open to the following problems: First, aluminum as a trivalent element binds three monoacid radicals, and therefore, the acid consumption is much larger than that of NaOH in alkali processes. Second, higher temperatures and energy consumption are needed for silicates to decompose and form aluminic salts than those for aluminate leaching. Third, Fe precipitation from an aluminic solution is a more difficult, power consuming process than desilication of an aluminate liquor. Fourth, the majority of alumina produced by acid process does not meet the impurities requirements content for aluminum smelting. Fifth, equipment made of ordinary steel or cast iron is intensively corroded by acids and aluminic salts unlike that by contact with the alkali and aluminate solutions. Consequently, the apparatus used for acidic treatment has to be protected against corrosion. This results in increased equipment costs. Sixth, the majority of acids aggravate air pollution due to their volatility. Currently, lime or limestone-soda sintering methods carried out at temperatures of 1000 to 1200°C remain feasible since they lead to mullite destruction. Likewise, the stream of coal ash and attendant materials through the sintering kiln can be decreased if silica is pre-extracted from the initial ash. The following are the advantages of this approach: (1) Up to 70-75% of silica is pre-extracted from ash which amounts to about 420 kg per 1 ton of the processed ash; (2) The silicon precipitated as silica is a chemical product with given properties like composition, purity, whiteness, color, fineness, specific surface, etc. Therefore, a wide assortment of silicate commodities can be manufactured. (3) The obtained concentrate is quite similar in properties and composition to nepheline concentrate and red mud from high-silicon bauxite processing into alumina and byproducts at some alumina plants in Russia and Kazakhstan. The method for ash processing is selected depending on its chemical and mineral composition. The latter is defined by the completeness of coal ash heat treatment, or in other words, by the extent to which the thermal decomposition of clay or kaolinite proceeds into metakaolinite, mullite, cristobalite or glassy matter. Increasing the combustion temperature for coal containing 10-12% of the sum of CaO and MgO favors the conversion of kaolinite into mullite and cristobalite. Therefore, ash obtained by the combustion of bituminous and subbituminous coals is more suitable for preliminary desilication. However, this must be taken as highly questionable since the degree of ash mullitization is unstable even at a single power plant. For this reason, knowledge of the mineral and chemical composition of the particular ash is a key factor in selecting a process. Technology for coal ash utilization for alumina production has not been previously developed and commercialized. Unfortunately, most of the researches carried out in this field, were interrupted in the 1970s after the world petroleum and bauxite markets improved. Nevertheless, there are three reasons necessitating continuing work on this problem. The first one is the environmental risk factor, because waste products from coal combustion have high potential risk associated with the effluent stream of metals in coals. Another reason is that over 50 % of the U.S. electric power is generated using coal. Coal deposits within the U.S. border represent one-third of the entire world's known supply. Consequently, ash landfill capacity has to be extended while requirements for handling and disposing of coal ash will become stricter. The third reason is that aluminum is expected to resume an annual growth pattern of 2 to 3 percent. On this basis, about 13 new alumina plants and aluminum smelters will be needed in the not very distant future. Coal ash should be considered for supplying the new plants with aluminum-bearing raw materials. The United States is the major world aluminum producer, but it needs to import about 90% of the aluminum-bearing raw materials. Therefore, coal ash as a domestic resource offers an advantage over foreign bauxite. The data obtained provide the basis for moving to on the next stage of process development. This stage should include design and operation of a chemical plant demonstration project with coal ash processing into diverse chemicals and construction materials.

References:

1.R.S. Kalyoncu, "Coal Combustion Products," U.S. Geological Survey, 2002 (www.minerals.er.usgs.gov).

2.Clean Coal Technology Roadmap, DOE, CURC, EPRI, 01/06/2004 (www.netl.doe.gov)

3.S.S. Tyson, T. Blackstock, "Coal Combustion Fly Ash," - Overview of Applications and Opportunities in the USA, Proceedings of the 211th ACS National Meeting, New Orleans, LA, March 24-28, 1996, p. 587-591.

4.D.M. Golden, "Research to Develop Coal Ash Uses," Proceedings of 9th International Ash Use Symposium, Orlando, FL, January 22-25, 1991, p. 69/1 - 69/23.

5."Coal Handbook," ed. By R.A. Meyer, Power and Environment; Dekker, New York, 1987.

6."Coal Quality and Ash Characteristics", A Study by the IEA Coal Industry, Advisory Board, OECO/IEA, 1985, p. 64. .

7.P. Bosh, T. Holzapfel, and I. Scholz, Zem.-Kalk-Gips, 1986, 39 (1), p. 36-38.

8.M. Patil, M. Sheraton, and M. Titelbaum, Fuel, 1984, 63 (6), p. 788-792.

9.A.N. Suturin, et al, "Complex Use of Urkutsk District Coal Ash," Research Report, Institute of Geochemistry of Suberian Division of Academy of Science of the USSR, Irkutsk, Russia, 1988.

10.E.P. Stambaugh and G.F. Jachsel, U.S. Patent 4,055,400, October 25, 1977.

11.E.P. Stambaugh and S.P. Chauhan, U.S. Patent 4,095,955, June 20, 1978.

12.E.P. Stambaugh, U.S. Patent 4,121,910, October 24, 1978.

13.L.C. McCandless and G.Y. Contos, "Current Status of Chemical Coal Cleaning Processes - an Overview", Proceedings of Symposium of Coal Cleaning, EPA, Washington, DC, 1978, v. 2.

14.R.A. Meyers, J.W. Hamersma, R.M. Baldwin, et al., "Low-Sulfur Coal Obtained by Chemical Desulfurization Followed by Liquefaction," Energy Sources, 1976, 31 (1), p. 13-18.

15."Coal Desulfurization," ed. By R.A. Meyers, M. Dekker, New York, 1977.

16.R.A. Meyers, L.J. Van Nice, and M.I. Santy, " Review and Status of the TRW/Meyers Process," Proceedings of the National Conference "Energy Environment Protection", 1976, 4, p. 372-385.

17.R.A. Meyers, "Compliance Coal for Utility, Industrial, Commercial, and Institutional Boilers via Gravichem Processing," Proceedings of an Annual Meeting of Air Pollution Control Association, 1978, 71 (2).

44

18. N.I. Eremin, S.S. Nurkeev, L.G. Romanov, et al., SU Patent 1,108,073, August 15, 1984.

19. S.S. Nurkeev, M.M. Kospanov, M.S. Tarasov, et al., SU Patent 247,346, July 30, 1986

20. S.S. Nurkeev, S.V. Burmistrov, L.G. Romanov, et al., SU Patent 908,748, February 28, 1982.

21. S.S. Nurkeev, E.A. Tastanov, and A.G. Morgunova, "Study of Sulfuric Acid Leaching of Ekibastuz Power Industry Slag," Metall. Obogashch., 1977, 12, p. 54-60.

22. A.P. Sabitov, S.S. Nurkeev, M.M. Kospanov, et al., "Optimization of Autoclave Decomposition of Power-Plant Slags of Ekibastuz Coal by Nitric Acid," J. Kompleksn. Ispol'z. Miner. Syr'ya, 1984, 5, p. 62-66.

23. L.A. Rith, "Combustion and Desulfurization of Coal in a Fluidized Bed of Limestone," Fluid Technology, Proceedings of Informative Fluid Conference, 1976, 2, p. 321-327.

24. G. Wargalla, J. Lotze, "Environmentally Safe Combustion of Low-Grade Coal in a Circulating Fluidized Bed," Erzmetall, 1985, 38 (5), p. 285-289.

25. K. Felker, F. Seeley, Z. Egan, and D. Kelmets, "Aluminum from Fly Ash," Chemitech, 1982, 12 (2), p. 123-128.

26. A.E. Torma, U.S. Patent 4,242,314, December 30, 1980.

27. A.E. Torma, "Extraction of Aluminum from Fly Ash," J. Metall.: Berlin, 1983, 37 (6), p. 589-592.

28. W.J. McDowell and F.G. Seeley, Ger. Offen. 2,929,295, February 28, 1980.

29. B. Lisowyi, U.S. Patent 4,567,026, January 28, 1986.

30. W.J. McDowell and F.G. Seeley, U.S. Patent 4,254,088, October 24, 1980.

31. P.M. Duhart, Ger. Offen. 2,257,521, June 07, 1973.

32. O. Dejan, V. Constantinesku, V. Toaje, et al. Rom. RO 78,101, January 30, 1982.

33. O. Dejan, M. Dejan, N. Petrescu, et al. Rom. RO 79,542, April 29, 1983.

34. O. Dejan, N. Petrescu, V. Constantinescu, et al. Rom. RO 76,831, July 30, 1981.

35. B. Lisowyi, D. Hitchcock, H. Epstein, U.S. Patent 4,678,647, July 07, 1987.

36. F. Nemeth, B. Szellem, F. Vaderna, et al. Hung Teljes HU 61,251, December 28, 1992.

37. W.R. Livingston, D.A. Rogers, R.J. Chapman, and N.T. Bailey, "The Use of Coal Spoils as Feed Materials for Alumina Recovery by Acid-Leaching Routes. 1. The Suitability and Variability of the Feed Materials," Hydrometallurgy, 1983, 10 (1), p. 79-96.

38. W.R. Livingston, D.A. Rogers, R.J. Chapman, et al. "The Use of Coal Spoils as Feed Materials for Alumina Recovery by Acidic-Leaching Routes. 2. The Effect of the Calcination Conditions on the Leaching Properties of the Colliery Spoil," Hydrometallurgy, 1983, 10 (1), p. 97-109.

39. W.R. Livingston, D.A. Rogers, R.J. Chapman, et al. ""The Use of Coal Spoils as Feed Materials for Alumina Recovery by Acidic-Leaching Routes. 3.The Effect of the LeachingConditions on the Extraction of Aluminum and Iron from a Fluidized Bed Ash," Hydrometallurgy, 1985, 13 (3), p. 283-291.

40. P. Mahi and N.T. Bailey, "The Use of Coal Spoils as Feed Materials for Alumina Recovery by Acidic-Leaching Routes. 4.The Extraction of Iron from Aluminoferrous Solutions with Amines, in Particular Alamine 336," Hydrometallurgy, 1985, 13 (3), p. 293-304.

41. G.O. Malibaeva, L.G. Romanov, S.S. Nurkeev, and A.G. Zaripova, "Autoclave Hydrochloric Acid Leaching of Ash from Low Temperature Combustion of Ekibastuz Coal for Alumina Recovery," J. Kompleksn. Ispol'z. Miner. Syr'ya, 1985, 6, p. 51-54.

42. G.O. Malibaeva and S.S. Nurkeev, "Hydrochloric Acid Leaching of Ekibastuz Coal Ash from the Ermakovsk Power Plant for the Recovery of Alumina," J. Kompleksn. Ispol'z. Miner. Syr'ya, 1986, 1, p. 45-49.

43. G.O. Malibaeva, L.G. Romanov, and S.S. Nurkeev, "Filtration of Hydrochloric Acid Slurries Produced by Leaching Ash of Ekibastuz Coal ," J. Kompleksn. Ispol'z. Miner. Syr'ya, 1985, 1, p. 57-61.

44. J. Russ, Jr.; J. Russ, Sr.; and R.T. Heagy; U.S. Patent 4,130,627, December 19, 1978.

45. Y. Sasaki, "Fundamental Tests on Recovery of Valuable Matter from Coal Ash. III.Chemical Treatment. 2.Calsinter Method," Chem. Abstr., 1989, 111 (54)/(23), p. 69-85.

46. C. Abbruzzese, G. Marzocchi, and G. Rinelli, "Possibility of Recovering Alumina from Ashes," Ind. Min., 1979, 30 (6), p. 323-328.

47. R.T. Hemmings, E.E. Berry, and D.M. Golden, "Direct Acid Leaching of Fly Ash: Recovery of Metals and the Use of Residues as Filters," Eighth International Coal Ash Utilization Symposium, Washington, DC, 1987, October 29-31, p. 38/A-1 - 38/A-15.

48. J. Kumamoto, N. Imanish, et al. "Recovery of Metal Oxides from Coal Ash," KOBELCO Technol. Rev., 1990, 7, p. 53-57.

49. J. Russ, Jr. and J.W. Smith, U.S. Patent 4,539,187, September 03, 1985.

50. J.N. Lehto, PCT Int. Appl. WO 89/04,811, June 01, 1989.

51. S.A. Shcherban and V.L. Rayzman, "Alkaline Technology of Coal Fly Ash Processing into Metallurgical and Silicate Chemical Products," Proceedings of 1995 International Ash Utilization Symposium, Lexington, KY, October 23-25, 1995.

52. I.Z. Pevzner, A.S. Dworkin, M.Ya. Fiterman, N.I. Eremin, and Ya.B. Rosen, "Mathematical Simulation of Aluminate Liquor Desilication," Tsvetn. Met., 1975, 7, p. 49-52.

53. R. Padilla and H.Y. Sohn, "Extraction of Alumina from Coal Wastes by the Lime-Soda Sinter Process," Light Met. (Warrendale, PA), 1982, p. 81-94.

54. R Padilla and H.Y. Sohn, "Kinetics of Sodium Aluminate Formation in the Production of Alumina from Coal Wastes by the Lime-Soda Sinter Process," Light. Met. (Warrendale, PA), 1985, p. 47-70.

55.R. Padilla and H.Y. Sohn, "Alumina from Coal Wastes by the Lime-Soda Sinter Process: Leaching and Desilication of Aluminate Solutions," Light Met. (Warrendale, PA), 1983, p.21-38.

56.R Padilla and H.Y. Sohn, "Kinetics and Alumina Yield in the Lime-Soda Sinter Process for Alumina from Coal Wastes," Metall. Trans. B, 1985, 16 B (2), p. 385-395.

57.J.R. Frederick, M.J. Murtha, and G. Burnet, "Coal Fly Ash a Potential Resource for Aluminum and Titanum," Proceedings of the Symposium of Chemical and Geochemical Aspects of Fossil Energy Extraction, Ann Arbor, MI, 1980, p. 107-127.

58.M.J. Murtha and G. Burnet, "Power Plant Fly Ash as a Source of Alumina," Proceedings of the Mineral Waste Utilization Symposium, 1978, 6, p.334-390.

59.M.J. Murtha and G. Burnet, "New Development in the Lime-Soda Sinter Process for Recovery of Alumina from Fly Ash," Report of Ames Lab., No 5-2, 1979, Ames, IA.

60.M.J. Murtha, U.S. Patent 4,397,822, August 09, 1983.

61.M.J. Murtha and G. Burnet, "New Development in the Lime-Soda Sinter Process for Recovery of Alumina from Fly Ash, " Proceedings of the Fifth International Ash Utilization Symposium, 1979, 10 (1), p. 68-84.

62.C. Abbruzzese, "Lime-Sintering Process for Non-Bauxite Ores, Including Fly Ashes," Trav. Com. Int. Etude Bauxites, Alumine, Alum, 1982, 17, p. 1-9.

63.J. Grzymek, "Connecting the Fabrication of Aluminum Oxide and Cement Starting from Materials Low in Aluminum Content by the Grzymek Method," Proceedings of the 4th International Congress Study Bauxites, Alumina, and Aluminum, 1978, 3, p. 74-95.

64.N.I. Eremin, G.Z. Nasyrov, P.S. Petel'ko, et al. "Processing of the Mineral Part of Ekibastuz Coal into Alumina and Cement by Sintering," J. Kompleksn. Ispol'z. Miner. Syr'ya, 1978, 50, p. 62-66.

65.J. Harsanyi, L. Kapolyi, P. Siklosi, and G. Vamos, Ger. Offen. 2,643,479, April 21, 1977.

66.B.J. Skifvars and M. Hupa, "Sintering of Ash During Fluidized Bed Combustion," Ind. Eng. Chem. Res., 1992, 31 (4), p. 1926-1030.

67.I.A. Semchenko, E.A. Rumyantseva, et al. SU Patent 922,092, April 2, 1982.

68.V.S. Sazhin, "New Hydrochemical Procedures for Complex Processing Aluminosilicates and High-Silicon Bauxites," Metallurgiya (Moscow), 1988, 215 p.

69.V.L. Rayzman, Yu.K. Vlasenko, M.M. Neusikhin, and I.Z. Pevzner, "Application of Hydrothermal Processes for Digestion of High-Silicon Aluminum-Containing Raw Material," Tsvetmetinformatsiya (Moscow), 1981, 60 p.

70.L.K. Hudson, "Aluminum from Coal Wastes," Technical Report of the Institute Mining Mineralogy Researches, Proceedings of Kentucky Coal Refuse Disposal Utilization Seminar, 1977, 3, p. 79-82.

71.A.B. Adrian, and H.W. McCulloch, "Pressure Leaching of Ores with Particular Reference to the Upgrading of Aluminate Solution. II.Alkaline Pressure Leaching of Sasol Fly Ash," Proceedings of Gov. Met. Lab. Of Rep. South Afr., 1966, 65.

72. L.G. Romanov, S.A. Shcherban, Kh.N. Nurmagambetov, and B.Sh. Dzhumabaev, "Extraction of Alumina from Ekibastuz Coal Ash by the Use of Hydrochemical Method," Proceedings of Institute of Metallurgy and Enrichment of Kazkh Academy of Sciences, 1966, 16, p. 118-121.

73. M. Pedlik and B. Hrncir, "Hydrothermal Alkaline Leaching of Aluminum-Containing Silicate Raw Materials," Freiberg Forschung, 1985, A 725, p. 19-35.

74. A.N. Zagorodnaya, S.A. Shcherban, Kh.N. Nurmagambetov, and V.D. Ponomaryov, "Crystallization of Sodium Hydroaluminate from Solutions Produced aftor Autoclave Treatment of the Ekibastuz Coal Ashes," Chem. Abstr., 1972, 69, 37746y.

75. I. Keil, P. Roessel, M. Hadan, et al. Ger. (East) DD 265,389, March 01, 1989. 76. N.G. Siliaieva, S.A. Shcherban, S.Kh. Tazhibaeva, and L.G. Romanov, "Investigation of Solubility of Ekibastuz Coal Ashes Components in Alkaline Solutions," J. Kompleksn. Ispol'z. Miner. Syr'ya, 1982, 3, p. 62-66.

77. V.L. Rayzman, L.P. Ni, S.A. Shcherban, et al. "Chemical Beneficiation of High-Silica Aluminum-Containing Raw Materials," Tsvetmetinformatsiya, Moscow, 1987, 60 p.

78. L.G. Romanov, E.F. Osipova, S.A. Shcherban, et al. "Alkaline Concentration of Ash Obtained by the Combustion of Ekibastuz Coals," Proceedings of Institute of Metallurgy and Enrichment of Kazakh Academy of Sciences, 1967, 23, p. 63-66.

79. S.A. Shcherban, Kh.N. Nurmagambetov, V.L. Rayzman, et al. "Processing of Ekibastuz Coal Ashes into Alumina and Byproducts," J. Kompleksn. Ispol'z. Miner. Syr'ya, 1979, 9, p. 71-75.

80. V.L. Rayzman and S.A. Shcherban, "Recovering Alumina, Silica and Byproducts from Coal Ash Through the Use of Process for Silicon Pre-Extraction," Light Met. (Warrendale, PA), 1997, p. 133-136.

81. S.A. Shcherban, "Alkaline Technologies of Coal Fly Ash Processing into Metallurgical and Silicate Chemicals Products," Proceedings of the International Ash UtilizationSymposium, 1997, October 23-25, Lexington, KY (Poster).

82. S.A. Shcherban and V.L. Rayzman, "Coal Ash Utilization by the New Chemical-Metallurgical Processes," Proceedings of the International Ash Utilization Symposium, 1997, October 20-22, Lexington, KY, p. 62-69.

83. J. Michelet, "Extraction de l'alumina a partir de shistes honillers: le procede H+. Mem. Et'etude. Sci. Rev. Met., 1982, 79 (1), p. 15-20.

84. C. Kirby, J. A. Barlay, "Alumina from Nonbauxite Resources" Trav. Com. Int'etude bauxites, alumine et alum., 1981, 16, p. 1-12.

85. K.B. Bengston, P. Chuberka, R.F. Nunn, et al. "Some Technical and Economic Comparisons of Six Processes for the Production of Alumina from Nonbauxite Ores", Trav. Com. Int'etude bauxites, alumine et alum., 1981, 16, p.109-132.

86. S.P. Kim, "Development of the Processes for Alumina Production from Low-grade Raw Materials Abroad", Bull. Tsvetnaya Metallurgia (Moscow: Tsvetmetinformatsia), 1987, 4, p. 91-93

87. K. Griotheim and B. Welch, "Impact of Alternative Processes for Aluminum Production on Energy Requirements," J. Met, 1981, 09, p.26-32.

88. I. Podushkova, V. Novotna, "Production of Al2O3 from Nonbauxite Raw Materials by the Alkaline Hydrometallurgical Method" (Budapesht: Bul. of Light Metals Production's Committtee, SEV), 1979, 12, p. 157-165.

89. Kh.N. Nurmagambetov, S.A. Shcherban, V.L.. Rayzman, et al. "Pilot Plant Tests of Complex Processing of the Ekibastuz Coal Ash into Alumina and Byproducts", Vopr. Proektir. I Ekspluatatsii Moshchn. Parogeneratorov na Ekibastuzsk. Uglyakh, Alma-Ata, USSR, 1976, p. 305-307.

90. Kh.N. Nurmagambetov, S.A.. Shcherban, G.B. Irgaliev, et al. "Pilot-Plant Testing of Ash Concentrate Processing by Sintering", Metall., Metallovedenie, Alma-Ata, USSR, 1977, 6, p. 58-66.

91. Kh.N. Nurmagambetov, S.A.. Shcherban, I.Z.. Pevzner, et al. "Pilot-Plant Testing of Hydrochemical Ash Cake Processing," Metall., Metallovedenie, Alma-Ata, USSR, 1977, 6, p. 67-73.

92. S. A.. Shcherban, V. L.. Rayzman, I. Z. Pevzner, "Technologies of Coal Fly Ash Processing into Metallurgical and Silicate Chemical Products", Proceedings of 210th ACS National Meeting, Chicago, IL, August 20-25, 1995, 40(4), p. 863-867.

93. "Technical and Economical Report about Utility for Processing of Ekibastuz Coal Ash into Alumina and Byproducts".All-Union Aluminium and Magnesium Institute (VAMI), Leningrad, USSR., 1979, 198 p. "Additional Considerations and Calculations", VAMI, Leningrad, USSR, 1980, 31p.

94. Yu.A. Lainer, "Comprehensive Processing of Aluminum-Containing Raw Materials by Acid Methods", Nauka, Moscow, 1982, 208 p.

2.2. Utilization of Lignite Ash by Alkaline Hydrothermal Method. Lignite is an imperfectly formed coal which fields are widespread in many continents and countries. Among all fossil fuels, lignite is distinguished by having the longest range in list reserves. On evidence deriverd from internet (*Rheinbaum Engineering and Wasser GmbH/RV Handel and Dienstleistungrn GmbH, 1997*), world lignite mining levels of 956 million metric tons (m t) from the commercially minable deposits can thus be maintained for 241 years. By comparison, the static range for oil and natural gas reserves is limited to several decades. Lignite is mined on all continents except for Africa. One centre of lignite mining is Western and Central Europe - unlike the situation in oil, natural gas and hard coal. In 1997, some 58 % of worldwide lignite output was accounted for by Central Europe. Other contributors are Russia with 9 %, North America with 12 %, Asian countries with 15 %, and Australia with 6 %. The largest lignite producers are Germany, Russia, the United States, Poland, Greece, Australia, the Czech Republic, China, Turkey and Canada. These ten countries' total output of 725 mt makes up three quarters of the world total. The main purchasers of lignite are power plants which are mostly located in the direct vicinity of mining areas. Taking into account that average ash content in lignite coal is 20% and Al_2O_3 percentage in the obtained ash is about 20%, it is easy to estimate that at least 29,000,000 tons of aluminium oxide, called alumina, being part of ash are burried. As to the U.S., amount of alumina contained in the dumped lignite coal ash is about 4,5 million metric tons. A great many chemical methods have been proposed for processing alumina-bearing coal ash [1]. Most of these processes are distinguished by the type of either of two acidic or alkaline extractants. The acidic-leaching route is unusable for decomposition of mullite that is a significant aluminum-containing ash constituent. Both basic ALHTP and Soda-Limestone Sintering Process (SLSP) are suitable for recovery alumina and byproducts from coal ash [2]. Examples of lignite ash composition are presented in Table 2.2.1.

50

Table 2.2.1. Examples of lignite ash composition

Oxide	SiO_2	Al_2O_3	Fe_2O_3	CaO	Na_2O	K_2O	SO_3	Total
Example 1(wt%)	44.0	23.0	14.0	6.5	3.0	1.7	1.5	93.7
mol/mol Al_2O_3	3.25	1.0	0.39	0.51	0.21	0.08	0.08	
Example 2(wt%)	51.0	19.5	17.0	2.0	3.0	1.8	1.5	95.8
mol/mol Al_2O_3	4.45	1.0	0.56	0.19	0.25	0.10	0.10	

Comparison of ALHTP with SLSP is performed by Example 1.

The foundation of classical SLSP is formation of sodium aluminate (SA) $NaAlO_2$ and dicalcium silicate Ca_2SiO_4 as a result of high temperature (1220-1400°C) solid-phase interactions between aluminosilicate like mullite, glassy phase, soda and limestone [3]. Simplified molar balance (SMB) of SLSP is diagrammatically shown in Figure 2.2.1.

The sinter obtained is leached by process water and slurry formed is separated into aluminate solution and Ca_2SiO_4-containing sludge. The latter is water washed and used for construction materials production (CMP). The solution is desilicated and then carbonized by cooled and cleaned flue gases. As a result, alumina hydrate $Al(OH)_3$ is precipitated and soda-sulfate solution (SSS) is depleted. $Al(OH)_3$ is calcined to alumina and SSS is evaporated to crystallize the reusable soda hydrate and to remove sulfate residue from the process. In contrary to SLSP, the classical ALHTP is based on formation of sodium-calcium hydrosilicate (SCHS) $NaCaHSiO_4$ and conversion of aluminosilicate into aluminate solution by raw material digestion. This procedure is carried out in the presence of caustic solution and lime at temperature 250°C and higher [4]. Simplified molar balance of classical ALHTP is shown in Figure 2.2.2.

51

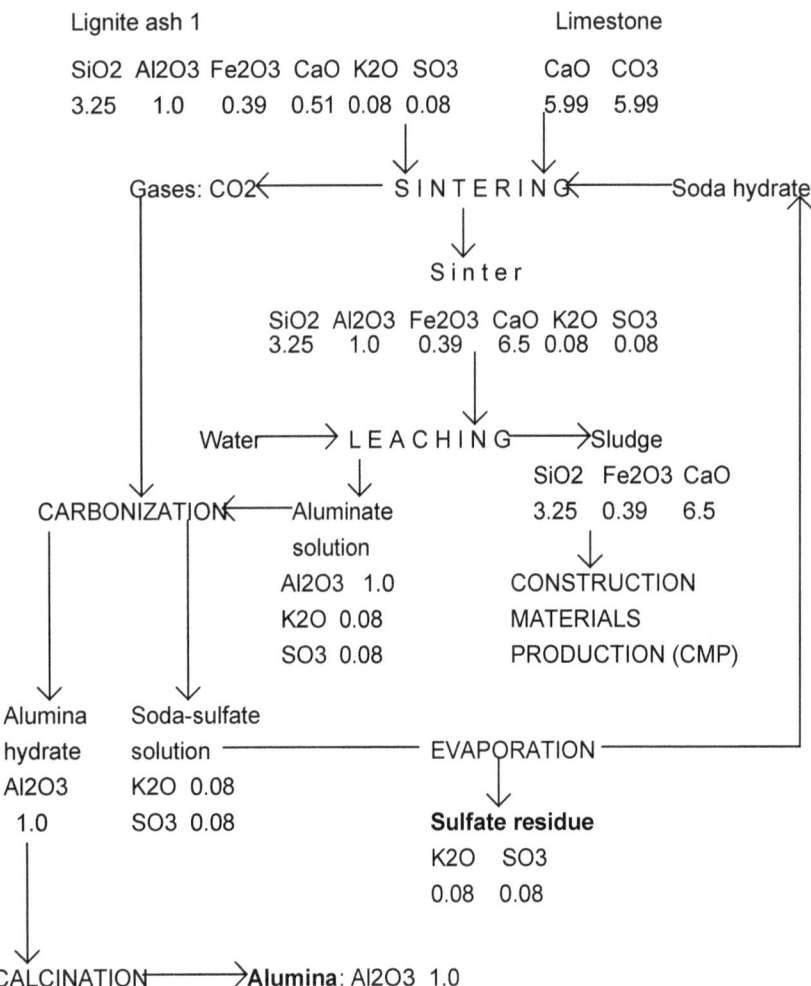

Figure 2.2.1.Simplified molar balance (SMB) of lignite ash utilization by SLSP

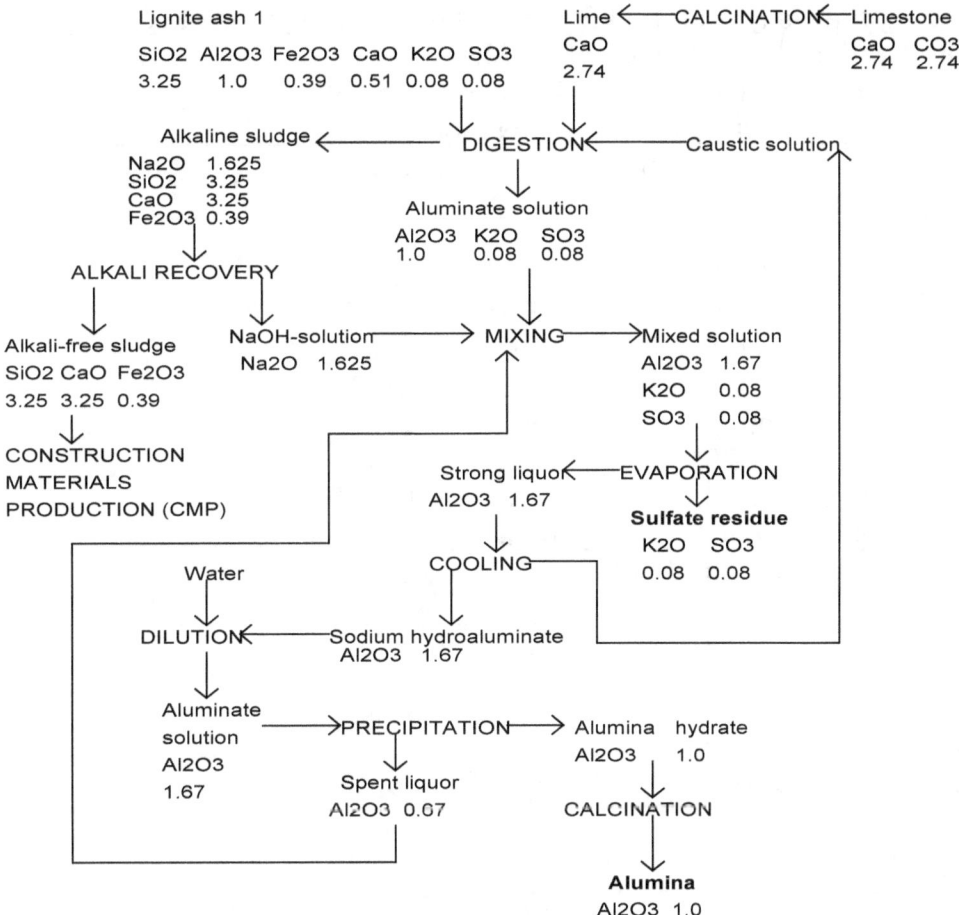

Figure 2.2.2. Simplified molar balance (SMB) of lignite ash utilization by the classical HTALP

Aluminate solution is consequently desilicated, evaporated to Na_2O concentration 550 g/l, then cooled and sodium hydroaluminate (SHA) $Na_2Al_2O_4 \cdot (2.5\text{-}3)H_2O$ is crystallized [5]. The crystals obtained can be used as a final chemical product or processed into $Al(OH)_3$. The sludge containing SCHS is treated with process water at 80-90°C giving alkali-free remainder and reused alkali liquor. Both ALHTP and SLSP compared ways were economically estimated. Coal ash [6] and other aluminosilicate-bearing raw materials (ASBRM) like anorthosite [7] and nepheline [4] have been considered. The data obtained are given in Table 2.2.2.

Table 2.2.2. Comparative HTALP and SLSP indexes with reference to different aluminosilicate-bearing raw materials (ASBRM)

Process	HTALP	HTALP	HTALP	SLSP
ASBRM	Coal ash	Anorthosite	Nepheline	
	Input per 1 ton Al_2O_3			
ton:				
ASBRM	4.47	4.38	3.96	3.93
Limestone	5.65	4.09	3.62	7.92
NaOH	0.36	n.s.*	0	0
Water	12.34	n.s.*	40.0	26.0
Power(kWh)	1500	n.s.*	600	1050
Fuel (GJ)	27.3	19.25**	36.8	60
($US)[8]	23.55	16.61	31.75	129.35
type	coal	coal	coal	oil
	Output per 1 ton Al_2O_3			
Caustic soda	0	0	0.97	0
Soda ash& potash	0	0	0	1.08
Calcium silicate-bearing sludge	7.56	6.49	4.66	5.94
Alumium-containing intermediate	TCHA***	SHA	SHA	SA

* not specified.
**Fuel input for SLSP as applied to anorthosite is reported to be 65 GJ/ton Al_2O_3 [7] that is estimated as $US 101.98 using gas and $US 140.13 using crude oil [8].
***Tricalcium hydroaluminate $Ca_3Al_2(OH)_{12}$

As is evident from the tabular data, principal streams are characterized by the bulk of feed materials and dumped sludges because of the high SiO_2 content (44-53%). Fuel inputs have the most impressive differences between ALHTP (19-37 GJ/ton Al_2O_3) and SLSP (60-65 GJ/ton Al_2O_3). These consumption items expressed in US dollars are as followed: $16.6-31.8 for ALHTP and $102-140 for SLSP. Correlation made by both thermal and monetary dimensions favours ALHTP. Its fuel consumption advantage is caused by the more perfect systems of transfering and recovery of heat as well as lower temperatures used for extraction of alumina. The type of fuel combusted, not its consumption only, determines the fuel cost. In this regard, preference should be also given to ALHTP that is based on indirect use of fuel gases heat for the feed stock decomposition. This procedure is carried out through steam generation as follows: fuel→FIRE BOX→fuel gases→BOILER→ steam→ HEAT EXCHANGER. Therefore, coal as the least expensive type of fossil fuel [8] can become the most profitable source of heat due to ALHTP commercialization. The contrastable sintering method predetermines fuel combustion directly in the sinterend blend. Therefore, mixing coal with raw material, limestone and soda leads to reducing percentage of combustible part of the prepared charge that becomes nonflammable.Consequently, only more expensive ash-free fuel like gas and oil is acceptable for SLSP heat supply. Lower cost of coal used for implementing ALHTP causes saving, in comparison with SLSP, at the minimum $70/ton Al_2O_3. The fuel consumption (FC) in ALHTP operation is close to that in the most efficienct Bayer process. In spite of obvious merits of ALHTP, its layout can be further improved with considerations of lignite ash composition features.

The Ways for ALHTP Improvement. Fe_2O_3 content in some lignite ashes is equal or higher than 14% (see Table 2.2.1). This feature makes possible to bind both iron and silicon in alkali-free calcium-ferrous hydrogarnet (CFHG) $Ca_3Fe_2(SiO_4)_m(OH)_{12-4m}$ [4,6]. Due to the CFHG formation, sodium percentage in the sludge is decreased and alkali recovery step of the process is reduced. It is reasonable to subject coal ash containing less than 3-5% CaO to preliminary silicon extraction (PSE) into caustic solution [3,9]. This procedure carried out at 90-100°C and Na_2O concentration 100-200 g/l allows to reduce the digested material stream removing up to 70% SiO_2. An obtained solid residue, so-called alumina-enriched concentrate, can be digested using double-flow layout that provides the separate heating of both feedstock and lime alkaline slurry for fouling mitigation[10]. Tricalcium hydroaluminate (TCHA) $Ca_3Al_2(OH)_{12}$ can be precipitated from aluminate solution instead of SHA at Na_2O concentrations about 200-300 g/l which is significantly lower than concentration required for SHA crystallization (550 g/l). In this case, Na_2O concentration gradient of the process is reduced and steam input decreased, though, heat consumption for calcination of extra limestone streams must be higher. This way was pilot-plant tested [6]. Coal ash ground with lime and caustic liquor was pumped through autoclave at the slurry rate of 30 l/hr. The digestion conditions were as followed: temperature 300°C, Na_2O concentration 300 g/l, residence time 10 minutes. 80% Al_2O_3 yield was obtained. As may be seen from Table 2.2.2 the limestone input 5.65 tons per 1 ton Al_2O_3 for this modification using TCHA precipitation exceeded the same index for the hydrothermal processing anorthosite (4.09 tons) and nepheline (3.62 tons). It was also asserted that ALHTP for ash utilization has been under development for about 10 years at Alcoa's pilot plant [12]. The best decision to be followed is found on the basis of raw material composition and local conditions. All innovations described above are embodied in Figure 2.2.3 covering SMB for ALHT-processing lignite ash of composition 2 low in calcium (CaO 2%) and high in iron (Fe_2O_3 17%).

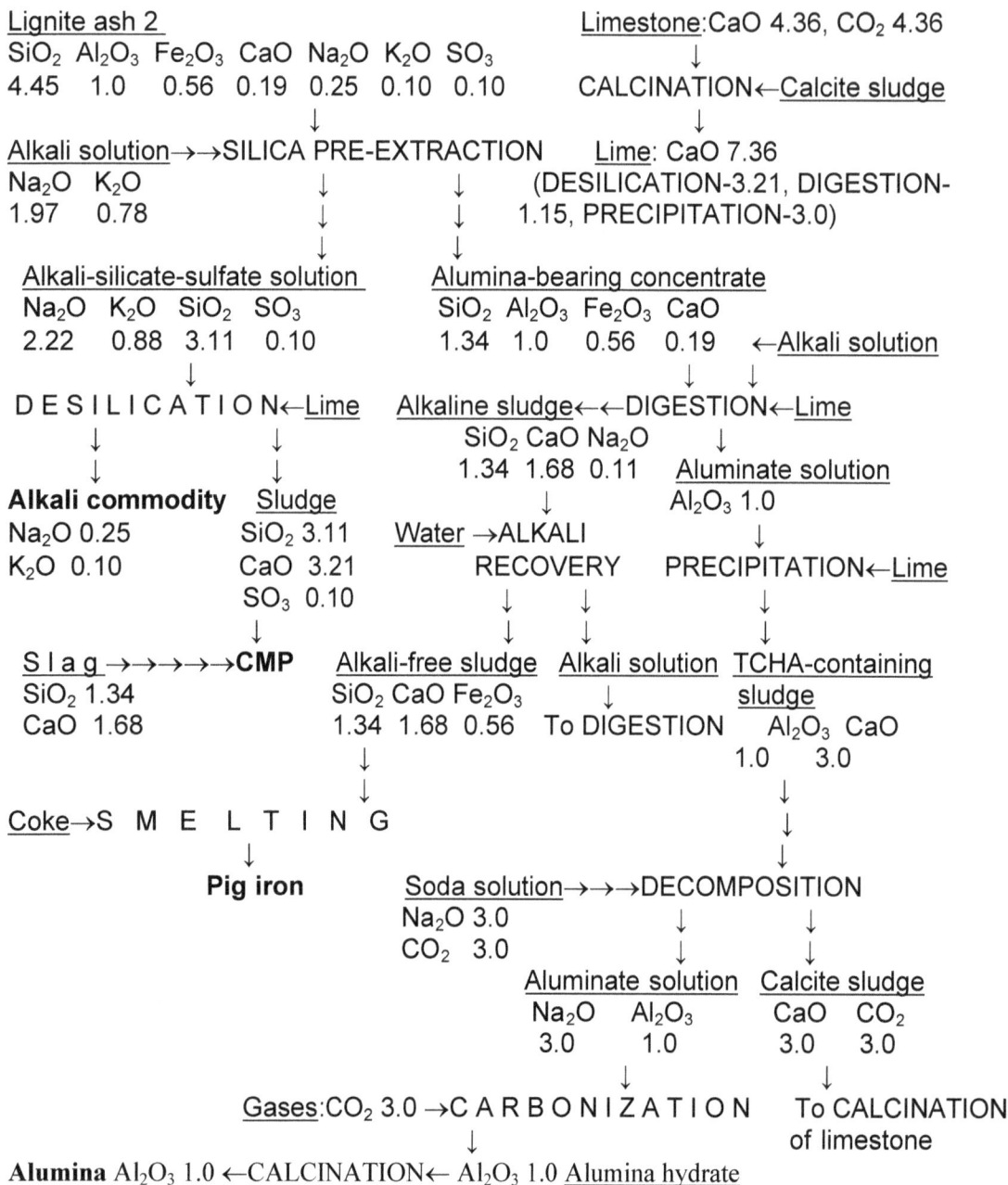

Lignite ash 2
SiO₂ Al₂O₃ Fe₂O₃ CaO Na₂O K₂O SO₃   Limestone:CaO 4.36, CO₂ 4.36
4.45   1.0   0.56   0.19 0.25 0.10 0.10   ↓
                             ↓   CALCINATION←Calcite sludge
Alkali solution→→SILICA PRE-EXTRACTION   Lime: CaO 7.36   ↓
Na₂O K₂O   ↓   ↓   (DESILICATION-3.21, DIGESTION-
1.97   0.78   ↓   ↓   1.15, PRECIPITATION-3.0)
                             ↓
     Alkali-silicate-sulfate solution   Alumina-bearing concentrate
     Na₂O K₂O SiO₂ SO₃   SiO₂ Al₂O₃ Fe₂O₃ CaO
     2.22   0.88 3.11   0.10   1.34 1.0   0.56   0.19 ←Alkali solution
                    ↓   ↓   ↓
D E S I L I C A T I O N←Lime   Alkaline sludge←←DIGESTION←Lime
     ↓   ↓   SiO₂ CaO Na₂O   ↓
     ↓   ↓   1.34 1.68 0.11   Aluminate solution
Alkali commodity Sludge   ↓   Al₂O₃ 1.0
Na₂O 0.25   SiO₂ 3.11   Water →ALKALI   ↓
K₂O 0.10   CaO 3.21   RECOVERY   PRECIPITATION←Lime
                    SO₃ 0.10   ↓   ↓   ↓
                     ↓   ↓   ↓   ↓
S l a g →→→→→**CMP** Alkali-free sludge Alkali solution TCHA-containing
SiO₂ 1.34   SiO₂ CaO Fe₂O₃   ↓   sludge
CaO 1.68   1.34 1.68 0.56   To DIGESTION   Al₂O₃ CaO
                              ↓   1.0   3.0
                              ↓   ↓
Coke→S M E L T I N G   ↓
           ↓   ↓
      Pig iron   Soda solution→→→DECOMPOSITION
                  Na₂O 3.0   ↓   ↓
                  CO₂   3.0   ↓   ↓
                                          ↓   ↓
                          Aluminate solution Calcite sludge
                          Na₂O   Al₂O₃   CaO   CO₂
                          3.0   1.0   3.0   3.0
                                      ↓   ↓
         Gases:CO₂ 3.0 →C A R B O N I Z A T I O N   To CALCINATION
                         ↓   of limestone
Alumina Al₂O₃ 1.0 ←CALCINATION← Al₂O₃ 1.0 Alumina hydrate

Figure 2.2.3. SMB of lignite ash utilization by the modified HTALP

56

The most complete pilot-plant tests of ALHTP extended over 30 years have been carrying out in Russia's National Research Institute of Aluminium and Magnesium (VAMI) [4, 6, 13]. The units for preparation, heating and digestion of raw material, flashing and filtration of slurry, washing cake, aluminate solutions desilication, evaporation and cooling, as well as SHA precipitation have been examined. Al_2O_3 yield in the course of hydrothermal process was so much as 88-92%.

Discussion and Conclusion. Application of ALHTP for lignite ash utilization can simultaneously solve some energetic, resources and environmental problems. The point is that lignite would be both raw material and fuel; in so doing ash dumping must lost any necessity. Total combination of these advantages is very important for many countries lacking commercial deposits of bauxite. Moreover, it was determined that the direct operating costs are almost the same for both plant based on ALHTP and alumina refinery processing bauxite by the Bayer method [15]. Beside of alumina hydrate and alumina, other valuable chemicals can be produced through ALHTP commercialization. Among these products are sodium hydroaluminate (SHA), tricalcium hydroaluminate (TCHA), silica-gel, calcite and calcium-ferrous hydrogarnet (CFHG). SHA is an effective coagulant for water purification as well as additive to mixtures for manufacturing drilling mud, waterproof concrete, zeolites, paper and cardboard [5]. The main TCHA field of usage is production of special high-alumina high impact, rapid-setting, straining and refractory cements [14]. As to CFHG, it should be processed into pig iron and calcium-silicate slag. Needless to say about handling $Al(OH)_3$ and Al_2O_3 since both have been worldwide processing into aluminium, refractories and burning retardants. The possibility to concentrate rare-earth elements in the remainder after double digestion of lignite ash must not be ruled out. ALHTP flexibility implies that its layout may be easily changed to turn out one or another product as the consumers' requirements. It should be noted that the lignite fired waste is a fine, loose and mining-free feedstock to give an additional advantage for its processing.

References:

1.V.L. Rayzman, S.A. Shcherban, and R.S. Dworkin, "Technology for Chemical-Metallurgical Coal Ash Utilization", Energy&Fuels, 11, No.4, 1997, p.761-773.

2.L.P. Ni and V.L. Rayzman, "Combining Methods for Processing Low-Grade Aluminum-Containing Raw Material", Alma-Ata, Nauka, 1988, 254p.

3."Production of Aluminum and Alumina", ed. A.R. Burkin, New York, John Wiley&Sons, Inc., 1987, 241p.

4.V.S. Sazhin, "New Hydrochemical Methods for Complex Processing Aluminosilicates and High-Silicon Bauxites", Moscow, Metallurgiya, 1988, 215 p.

5.V. Rayzman, I. Filipovich, L. Nisse, and Yu. Vlasenko, "Sodium Aluminate from Alumina-Bearing Intermediates and Wastes", JOM, 50 (11), 1998, p.32-37.

6.I. Podushkova and V. Novotna, "Production of Al2O3 from Nonbauxite Raw Material by Hydrometallurgical Method", Bulletin of Committee of Light Metals Production, SEV, Budapest, 12, 1979, p.157-165.

7.K. Griotheim and B. Welch, "Impact of Alternative Processes for Aluminum Production on Energy Requirements", Journal of Metals, 9, 1981, p. 26-32.

8.Statistical Abstracts of the U.S. Bureau of the Census, Washington DC, 1996, 1024 p.

9.V.L. Rayzman and S.A. Shcherban, "Recovering Alumina, Silica and Byproducts from Coal Ash Through the Use of Process for Silicon Pre-Extraction", Light Metals, TMS/AIME Proceedings, Warrendale (PA, USA), 1997, p.133-136.

10.V.L. Rayzman, "Mass Transfer Considerations During the Precipitation of Deposits on Alumina Refinery Heat Exchanger Surfaces", Light Metals, TMS/AIME Proceedings, Warrendale (PA, USA), 1996, p. 5-10.

11.V.L. Rayzman, "Red Mud Revisited - Special Paper on Scandium Potential", Aluminium Today, 10 (5), 1998, p. 64-68.

12.H.J. Hittner, "Hydrothermal Alkaline Process to Extract Alumina from Anorthosite", Trav. Com. Int. Etude Bauxites, Alumine, Alum., Paris, 1981, 16, p. 13-21.

13.I.Z. Pevzner and V.L. Rayzman, "Autoclave Procedures in Alumina Production", Moscow, Metallurgiya, 1983, 128 p.

14.V.M. Sizyakov, V.I. Korneev, and V.V. Andreev, "Upgrading of Quality of Alumina and Byproducts Processed from Nephelines", Moscow, Metallurgiya, 1986, 118 p.

2.3. Heat Exchanger Fouling in Alumina Refinery as a Mass Transfer Phenomenon.

All mass transfer equipment of alumina refineries becomes encrusted with deposits. The dense fouling that is formed on the heat-exchanger surfaces causes an additional thermal resistance and sharply decreases the rate of heat transfer from the heat carrier to the heated fluid. The latter can be either an evaporated aluminate solution or leached raw material slurry. Lower heat-transfer coefficients bring about ineffective heat flux and fuel overconsumption. A reduction of deposition on heat exchanger surfaces, for example, by 50% would make it possible to reduce steam consumption by about 20%. It is important to understand what is behind exchanger fouling. Several empirical formulas have been proposed relating fouling rate with temperature (T) and properties of the heated fluid. For example [1]:

$$U_f = -582.97 + 6.63\,T - 0.0016\,T^2$$

where U_f = fouling rate from heating the bauxite slurry.

However, so far there has not been a satisfactory explanation of this phenomenon's driving conditions, although it is reasonable to approach the study of fouling formation kinetics from the standpoint of mass transfer theory.

Kinetics of Heat Exchanger Fouling in Alumina Manufacture. Scale formation on the heating surface can be considered as the result of a heterogeneous chemical reaction that occurs from a triple collision, with one of the collision participants being the material of the heat transferring wall. The collision is followed by the type of bimolecular chemical reaction A + B > products [2].

The rate constant **k** of this interaction is expressed symbolically, as follows:

$$\ln(\mathbf{k}) = \ln(PL) - (E/R)T^{-1} + c\ln T$$

where P = steric factor, reflecting the spartial arrangement of active centers,

L = collision factor,

E = activation energy,

R = gas constant, T = surrounding temperature.

It can be assumed that the fouling rate U_f is directly proportional to the constant **k** and that the hydrodynamic, physical, and chemical parameters of the heated fluid are invariate (E=const). Then:

$$\ln U_f = K[\ln U_{lm} - (E/R)T^{-1} \pm c\ln T] \qquad (1)$$

where K = coefficient of proportionality.

The terms of Equation (1) within the parenthesis reflect various aspects of the scale deposition process, *viz:*

U_{lm} = PL = linear mass-transfer factor,

(E/R)T-1 = factor of crystal seed formation,

± clnT = factor of crytstallized deposit solubility.

In its turn [3]:

$$U_{lm} = U_m/\rho = \beta(c_b - c_L) \qquad (2)$$

where U_m = rate of substance mass transfer,

ρ = solid substance density,

β = mass transfer coefficient,

c = substance concentration,

b and L = subscripts denoting the bulk of the heated fluid and boundary layer adjoining the phase interface surface.

According to the law of diffusion coefficient decay, close to phase interface surfaces [4]

$$Nu' = 0.016\ Re^{7/8}\ Sc^{1/3}$$

where

Nu' = duffusion Nusselt number,

Re = Reynolds number,

Sc = Diffusion Schmidt number

By using the familiar formulas

$$Nu' = \beta d/D;\ Re = (U_v d)/\vartheta;\ Sc = \vartheta/D$$

where

D = diffusion coefficient,

d = diameter,

U_v = flow velocity,

ϑ = kinematic viscosity coefficient

After some transformation, the result can be expressed as

$$\beta = 0.016\ (D^{2/3} / \vartheta^{13/24})\ U_v^{0.875} d^{-0.125}$$

Thus, the complete extended form of Equation (1) becomes:

$$lnU_f = K/a(c_b - c_L)\ ln(U_v^{0.875} d^{-0.125}) - (E/R)T^{-1} \pm c\ lnT/ \qquad (3)$$

where a = coefficient of proportionality

Pilot Plant Test of Theoretical Predictions. In order to determine the interaction of all hydrodynamic quantities with respect to scale deposit formation, a pilot heat exchanger was designed and built [5]. The above-mentioned apparatus was related to the type of heat exchanger "tube-in-tube". The tubes were arranged vertically. The heated solution was pumped through internal tube of diameter ***d***. Water saturated steam was obtained from a power plant using an internal tube having a diameter of 50 mm and length ***L***. Dimensions of the heat exchanger were varied within the following limits: ***d*** from 4 to 19 mm, ***L*** from 2.3 to 10.5 meters. Flow velocity U_v was varied from 0.19 to 1.26 m/s. The actual spent process solution t was heated in the heat exchanger. Similarly to the method described in Reference [6], additions of sodium silicate (water glass, in this case) were used to reach a sufficient level of liquor saturation by SiO_2 to obtain the desired quantity of precipitated desilication product (DSP). The composition of initial aluminate-silicate spent solution was as follows: $Na_2O_{caustic}$ 140-150 g/l, Al_2O_3 62-75 g/l, SiO_2 3 g/l. The temperature of incoming and discharged solutions was maintained constant at 30°C and 90°C, respectively. Thus, the physical, thermotechnical and chemical conditions in this pilot plant test were constant. Every experiment was started with a clean heat transfer surface and considered to be terminated after the sharp fall in heat-transfer coefficient. The encrusted heating tube was removed and sawed into ten pieces. The fouling thickens from both the first and second cuts of each piece was measured, and an average fouling size and rate were calculated. The results are shown in Table 2.3.1.

Table 2.3.1.Data of DSP deposition in the tube-in-tube pilot plant exchanger

Internal clean tube diameter d (m) x 10^3	Flow velocity U_v (m/s)	Expression $\ln(U_v^{0.875}d^{-0.125})$	Fouling rate U_f (m/s)x10^9	Parameter $\ln U_f$
9	1.26	0.791	53	-16.753
9	1	0.589	62	-16.596
4	1	0.69	57	-16.68
9	0.89	0.487	68	-16.503
4	0.69	0.366	78	-16.366
14	0.51	-0.056	107	-16.05
19	0.45	-0.203	121	-15.927
9	0.39	-0.235	125	-15.895
14	0.28	-0.58	167	-15.605
11	0.26	-0.615	172	-15.576
9	0.19	-0.864	212	-15.366

As it is shown from Figure 3.1.1 the tabulated data are described by a straight line equation, as follows:

$$\ln U_f = -0.842 \ln (U_v^{0.875} \cdot d^{-0.125}) - 16.095 \qquad (4)$$

lnU$_f$

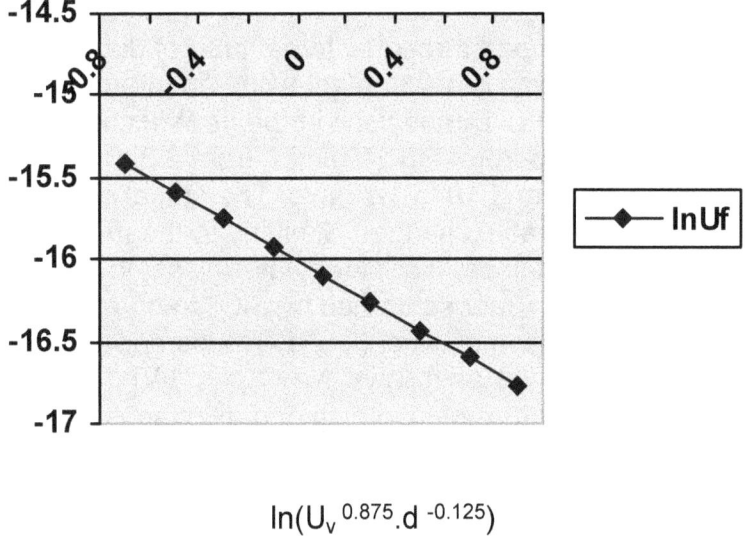

$$\ln(U_v{}^{0.875}.d^{-0.125})$$

Figure 2.3.1. Fouling rate U$_f$ as a function of the flow velocity U$_v$ and internal tube diameter d

The suitability and reliability of the obtained theoretical and practical results were checked by calculating the known data of heating some bauxites and nepheline slurries [7-10]. The function in (U$_f$.10^9) = A - BT-1 - ClnT expressed according to these data and Equation (4) are shown in Table 2.3.2.

The similarity of the obtained equations to Equations (3) and (4) supports the validity of the above calculations and conclusions. It is important to note that various types of heat exchangers and heat carriers were used for the above tests. These differences are reflected in the coefficients shown in Table 2.3.2.

Table 2.3.2. Results for calculating the parameters of an equation $\ln(U_f \times 10^9) = A - BT^{-1} - C\ln T$ on the strength of pilot plant tests data [7-10]

# Raw material	Heat exchanger type	Heat carrier	Slurry temperature (°C)		Main rock forming minerals Coefficients		
			Intake	Discharge	A	B×10³	-C
1. North Russian bauxite	Shell-and-tube	Diphenyl mix (DM) steam	85	260	Boehmite, gibbsite, kaolinite [7]		
					3.566	0.37	-0.311
2. Greek bauxite	Tube-in-tube	DM liquor	130	260	Diaspore [8]		
					954.98	65.31	131.87
3. Hungarian bauxite	Tube-in-tube	DM liquor	130	260	Boehmite [8]		
					773.61	51.55	107.14
4. Yugoslavian bauxite	Tube-in-tube	DM liquor	130	250	Gibbsite, boehmite [8]		
					137.79	10.37	17.989
5. Ukranian nepheline	Tube-in-tube	Water liquor	140	290	Albite, microkline [9]		
					1275.1	90.82	174.68
6. Turkish bauxite	Shell-and-tube	Water steam	140	170	Boehmite, diaspore [10]		
					13.514	5.712	-0.48

Application of the Test Results. Equation (3) predicts the same value of fouling rate for both pilot and commercial heat exchangers if coefficients **K**, **a**, and **c**, as well as the concentration difference (C_b - C_L) and temperature **T**, are the same for both types of installations. Therefore, for obtaining similar heat exchanger fouling rates, the following equality should be satisfied:

$$(U_v^{0.875} \cdot d^{-0.125})_{pp} = (U_v^{0.875} \cdot d^{-0.125})_{cp}$$

where the subscripts pp and cp refer to the pilot plant and commercial one, respectively.

Since the terms **d** and U_v for the heated Turkish bauxite slurry at the pilot plant were equal to 0.014 m and 1 m/s, respectively, the same parameters for the designed commercial Turkish Seidishekhir alumina plant d_{cp} and U_{vcp} have been selected, proceeding from the formula:

$$U_{vcp} = [1^{0.875} \cdot (0.32)^{0.125}/(0.014)^{0.125}]^{1/0.875} = 1.125 \text{ m/s}$$

where 0.032 is the internal diameter of the commercial heat exchanger pipe in meters.

As a result, the same fouling rate was obtained for both pilot and commercial heat exchangers at a slurry temperature of 170°C, viz: from 0.9 to 1.0 mm for 30-33 days of operation [10].

The Most Efficient Way for Mitigating Heat Exchanger Fouling. Equation (3) predicts that mitigating the deposition of precipitates on heat exchanger surfaces can be obtained by a reduction of flow velocity, an increase in heating pipe diameter, changes in fluid or lowering of the concentration difference. However, it is impractical to bring the flow velocity down to zero, to greatly increase the diameter of heating pipes, or to significantly change the process temperature. Therefore, decrease in the only practical approach to reduce heat exchanger fouling. This means that prior treatment of the heated fluid is required in order to precipitate the fouling constituent before it reaches the heat exchanger. An example of such pretreatment is desilication of aluminate solution before its heating [11].

References:

1.N.S. Mal'ts, "The Innovations in Alumina Production by the Bayer-Sintering Process," Moscow: Metallurgiya, 1989, 176 p.

2.T.L. Brown and G.E. LeMay, Jr., "Chemistry: The Central Science," Englewood Cliffs, NJ: Prentice-Hall, Inc., 1977, 816 p.

3.A.N. Planovsky, V.M. Ramm, and S.Z. Kagan, "Processes and Apparatus for Chemical Engineering, Moscow, Goskhimizdat, 1962, 846 p.

4.G. Astarita, "Mass Transfer with Chemical Reaction," New York, Elsevier Publishing Company, 1967, 224 p.

5.V.L. Rayzman, I.Z. Pevzner, N.S. Mal'ts, and M.M. Neusikhin, "Pilot Plant Study of Hydrodynamics Quantities Influence on Rate of DSP Deposition on the Heat Exchanger Surface," Moscow, Tsvetmetinformatsiya, 10, 1971, p. 37-39.

6.H. Muller-Steinhagen, M. Jamialahmad, and B. Robson, "Understanding of Mitigating Heat Exchanger Fouling in Bauxite Refineries," JOM, 11, 1994, p. 36-41.

7.V.L. Rayzman, M.M. Neusikhin, and M.Ya. Yakovlev, "Pilot Plant Tests of Indirect Heating of Bauxite Slurries," Chemical Abstracts, v. 102, 1985, # 27470q.

8.V.L. Rayzman, "The Kinetics of the Deposition of Precipitates on the Heat Exchanger Surface in Alumina Refining," Chemical Abstracts, v. 102, 1985, # 188534d.

9.V.S. Sazhin, V.L. Rayzman, V.G. Kazakov, et al., "Study of the Heating of Lime-Alkali Slurries of the Pilot Plant Autoclave Hydrochemical Processing of Mariupolites," Chemical Abstracts, v. 100, 1984, # 194438a.

10.V.L. Rayzman, "Modeling of Heat and Mass Transfer in Tubular Steam-Slurry Preheaters for Alumina Production," Chemical Abstracts, v. 100, 1984, # 36390z. 11.I.Z. Pevzner and V.L. Rayzman, "Autoclave Processes in Alumina Production," Moscow, Metallurgiya, 1983, 128 p.

2.4.Regularity of Hydrothermal Chemical Reactions*

There is a category of chemical interactions taking place at temperature above the boiling point of the liquid phase involved, as well as upon high-than atmospheric pressure. This group of chemical phenomenons is known as Hydrothermal Reactions (HTR). They are a great concern in both formation of the underground mineral deposits and metallurgy and chemistry where we deal with autoclave technique (digestion). The sufficient temperature T_{suf} of HTR is the key-factor causing magnitude of othe parameters of the procedure discussed. Determination of T_{suf} is resulted from a host of the thorough longterm experiments since each specific raw material requires a new set of investigation. The T_{suf} values are selected by diagrammatic interpolation and hence are generally approximate. An example of such graphic fixing of T_{suf} is shown in Figure 2.4.1 depicting the kinetics of boehmite digestion with alkali-aluminate solution /1/.

Al_2O_3 yield, %

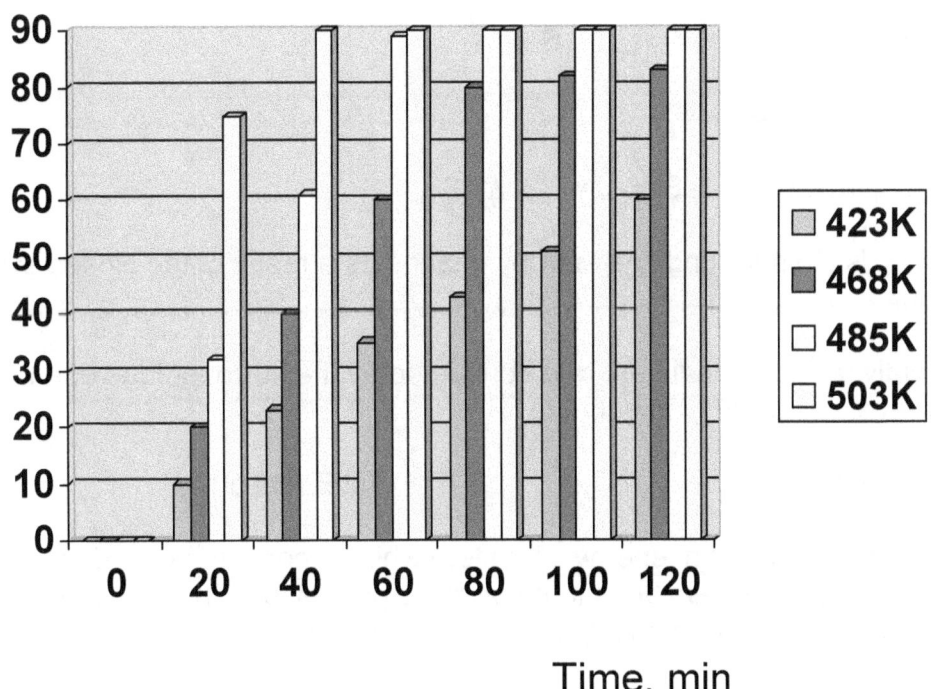

Time, min

Figure 2.4.1. Kinetics of boehmite digestion at various temperatures

*-Rayzman, V.L.; Interconnection of Entropy and Temperature of Reversible Autoclave Processes, Non-Ferrous Metals, 1983, #8, p. 45-48

As evident from diagram, the T_{suf} value is higher than the approaching magnitude 468K providing 85% of alumina extraction and less than or equal to 485K giving 95% of alumina yield. It should be stressed that the temperature value can be raised for speeding up the digestion procedure. As applied to boehmite treatment, the temperature increased from 485K (T_{suf}) to the excessive magnitude 503K (T_{exc}) allows to cut the digestion time from 60 minutes to 30 minutes. Whatever, the temperanure will be used, it should not be lower than T_{suf} level determined as a basis point. By this way, T_{suf} for the boehmite digestion is found equal to 485K. However, there is not a total assurance that the above value is absolutely true. It is desirable to calculate this parameter theoretically using laws and formulas of physical chemistry, in particular, chemical thermodynamics.

2.4.1. Development. Let us consider a closed system which has two states – initial (1) and final (2). Conversion of first state to the second may occur by both reversible and irreversible ways. As it follows from the properties, the total ($\Delta_T S_m$) and surroundings ($\Delta_T S^{rev}_{surr}$) entropy change parameters is equal to zero ($\Delta_T S_m + \Delta_T S^{rev}_{surr} = 0$). As to irreversible process, its total entropy grows, hence $\Delta_T S_m + \Delta_T S^{rev}_{surr} > 0$. A heat stream between the system and surroundings (Q^{rev} and Q^{irr}, respectively) is expressed by the surroundings entropy change and temperature T_{surr} as follows:

$$Q^{rev} = T_{surr} \cdot \Delta_T S^{rev}_{surr} \quad \text{and} \quad Q^{irr} = T_{surr} \cdot \Delta_T S^{irr}_{surr}$$

From this it follows that

$$Q^{irr} - Q^{rev} = T_{surr} \cdot (\Delta_T S^{irr}_{surr} - \Delta_T S^{rev}_{surr}) = T_{surr} \cdot \Delta_T S_{surr}$$

where ΔT_{surr} is the surroundings entropy change as a result of the closed system conversion.

Consequently, an additional heat rate consumed by the surroundings due to the process irreversibility Q^{irr}_{add} is equal to

$$Q^{irr}_{add} = T_{surr} \cdot \Delta_T S_{surr} \qquad \text{(1st Equation)}$$

It is evident that the internal energy of system while in conversion 1→2 is changed alike in both reversible and irreversible processes. This losses of resulting work (heat stream) taking place on account of irreversibility ΔW^{irr} is

$$\Delta W^{irr} = Q^{irr} - Q^{rev} \qquad \text{(2nd Equation)}$$

In the case when the heat is produced, $\Delta W^{irr} > 0$ and if the heat is spent, $\Delta W^{irr} < 0$ at the same absolute value. Correlating both 1st and 2nd Equations, we derive Gouy-Stodola formula /2/

$$\Delta W^{irr} = T_{surr} \cdot /\Delta_T S_{surr}/ \qquad \text{(3rd Equation)}$$

66

where $/\Delta_T S_{surr}/$ comprises an absolute magnitude of the surroundings entropy flux gain or loss as a result of substitution of reversible process for irreversible one. The Gouy-Stodola law is valid not only to the heat exchange process but also to chemical conversion in the closed system /3/. In chemical process parameter
ΔW^{irr} relates to increase (loss) of the work spent to a substance production. Let us assume that an ambient temperature is close to 298.15K. Such approximation introduces slight error into the calculations but enables to make use of the thermodynamic tables data. Then 3rd Equation becomes

$$\Delta W^{irr} = 298.15 \cdot /\Delta_f S_m/ \qquad \text{(4}^{th}\text{ Equation)}$$

where $/\Delta_f S_m/$ is the absolute value of standard entropy change of the chemical reaction in closed system, that is HTR.

In its turn, parameter ΔW^{irr} summarizes the work change resulting the chemical conversion carried out of closed system in a distinctive temperature range between T_1 and T_2, therewith T_1 defines the onset of HTR irreversibility and $T_2 = T_{suf}$. The corresponding reaction is featured by entropy change beginning with $\Delta_{T1} S^0_m$ and ending with $\Delta_{T2} S^0_m$. It is apparent that by the time of HTR completion at temperature T_{suf}, the entropy change will correlate with parameter ΔW^{irr}. This correlation is reflected by an equality

$$\Delta W^{irr} = 298.15 \cdot /\Delta_f S^0_m/ = T_{suf} \cdot /(\Delta_{T2} S^0_m - \Delta_{T1} S^0_m)/ \qquad \text{(5}^{th}\text{ Equation)}$$

The hydrothermal interactions under consideration are classified as isobaric (p=const) because pressure has not essential influence on the system constitution. The entropy change of HTR at temperature T is equal to

$$\Delta_{T2} S^0_m = \Delta_{T1} S^0_m + \int_{T_1}^{T_2} (\Delta C_p / T) dT$$

where ΔC_p is molar heat capacity change upon constant pressure.
Taking ΔC_p = const in the temperature range between T_1 and T_{suf} (so-called 2nd approximation), we can write

$$/(\Delta_{T2} S^0_m - \Delta_{T1} S^0_m)/ = \Delta C_p \cdot \ln(T_{suf}/T_1)$$

Inserting this expression in 5th Equation, the following equation would be derived

$$T_{suf} \cdot \Delta C_p \cdot \ln(T_{suf}/T_1) = 298.15 \cdot /\Delta_f S^0_m/$$

Hence it follows that

67

$/\Delta_f S^0_m/ = K \cdot T_{suf} \cdot \ln(T_{suf}/T_1)$ (6th Equation)
where $K = \Delta C_p/298.15 = const$

The numerical values of the coefficients of 6th Equation should be found by selection and correlation of thermodynamic and practical data.

2.4.2.Results. There are known HTR conducted at the established operating temperatures. These temperatures, as a rule, are placed between T_{suf} and T_{exc} values since researchers and producers usually tend to raise temperature for keeping the industrial procedure time down. Therefore, T_{suf} level should be expected to fall in the above mentioned temperature range. There were so much sources studied to collect the sought HTR /1, 4-22/. The reactions picked and their temperature ranges recommended are presented in Table 2.4.1.

Table 2.4.1.Equations and temperature ranges of the hydrothermal reactions

No	Equation of reaction	T,K
1) $\beta\text{-}Ca_2SiO_{4(s)}$ (larnite) + $H_2O_{(l)} = Ca_2SiO_4 \cdot H_2O_{(s)}$ (dicalcium silicate hydrate)		403-408 /4/
2) $Al_2Si_2O_5(OH)_{4(s)}$ (kaolinite) + $6H^+_{(aq)} = 2Al^{3+}_{(aq)} + 2SiO_{2(s)} + 5H_2O_{(l)}$		403-413 /5/
3) $Ni^{2+}_{(aq)} + H_2S_{(g)} = NiS_{(s)}$ (millerite) + $2H^+_{(aq)}$		403-413 /6/
4) $Ni^{2+}_{(aq)} + \alpha\text{-}FeS_{(s)}$ (troilite) $= NiS_{(s)} + Fe^{2+}_{(aq)}$		403-423 /7/
5) $CuFeS_{2(s)}$ (chalcopyrite) + $2H^+_{(aq)} = CuS_{(s)}$ (covellite) + $Fe^{2+}_{(aq)} + H_2S_{(g)}$		423-433 /7/
6) $Sb_2O_{5(s)} + HS^-_{(aq)} = 3OH^-_{(aq)} = 2SbO^+_{(aq)} + SO_4^{2-}_{(aq)} + 2H_2O_{(l)}$		403-433 /6/
7) $Al(OH)_{3(s)}$ (gibbsite) + $OH^-_{(aq)} = AlO(OH)^{2-}_{(aq)} + H_2O_{(l)}$		423-438 /1/
8) $CaO_{(s)} + 2Al(OH)_{3(s)} + 6H_2O_{(l)} = Ca_3Al_2(OH)_{12(s)}$ (calcium hydroaluminate)		433-448 /4/
9) $\beta\text{-}LiAlSi_2O_{6(s)}$ (β-spodumene) + $2Ca(OH)_{2(s)} + H_2O_{(l)} = Li^+_{(aq)} + OH^-_{(aq)} + \gamma\text{-}AlOOH_{(s)}$ (boehmite) + $2CaH_2SiO_{4(s)}$ (calcium hydrosilicate)		443-453 /8/
10) $3CaO_{(s)} + 2SiO_{2(s)} + 3H_2O_{(l)} = Ca_3Si_2O_7 \cdot 3H_2O_{(s)}$ (tricalcium disilicate trihydrate)		453-463 /4/
11) $CaO \cdot Al_2O_3 \cdot 4SiO_2 \cdot 2H_2O_{(s)}$ (calcium-aluminum tetrasilicate dihydrate) + $3H4SiO4(aq)$ (silicic acid) = $CaO \cdot Al_2O_3 \cdot 7SiO_2 \cdot 7H_2O_{(s)}$ (Ca-Al polysilicate polyhydrate, CAPSPH) + $H2O(l)$		463-473 /9 /
12) $2\beta\text{-}LiAlSi_2O_{6(s)} + Na_2CO_{3(aq)} + 2H_2O_{(l)} = Li_2CO_{3(s)} + 2NaAlSi_2O_{6(s)}$ (analcime)		463-473 /8/
13) $2CaO_{(s)} + 2SiO_{2(s)} + 2H_2O_{(l)} = Ca_2SiO_4 \cdot 2H_2O_{(s)}$ (dicalcium silicate dihydrate)		458-473 /4/
14) $2NaAlSi_2O6 \cdot H_2O_{(s)} + Ca(OH)_{2(aq)} = 2NaOH_{(aq)} + CaO \cdot Al_2O_3 \cdot 4SiO_2 \cdot 2H_2O_{(s)}$ (calcium-aluminum tetrasilicate dihydrate)		463-478 /9/

Table 2.4.1(continuation)

No	Equation of reaction	T,K
15)	$Ca(AlSi_2O_6)_2 \cdot 2H_2O_{(s)}$ (CAPSPH) + $2H_2O_{(l)}$ = $Ca(AlSi_2O_6)_2 \cdot 4H_2O_{(s)}$ (CAPSPH)	463-483 /9/
16)	$MnWO_{4(s)}$ (huebnerite) + $CO_3^{2-}{}_{(aq)}$ = $WO_4^{2-}{}_{(aq)}$ (tungstate Ion) + $MnCO_{3(s)}$ (rhodochrosite)	463-473 /10/
17)	$Pb^{2+}{}_{(aq)}$ + $S^{2-}{}_{(aq)}$ = $PbS_{(s)}$ (galenite)	463-479 /11/
18)	$CaWO_{4(s)}$ (scheelite) + $CO_3^{2-}{}_{(aq)}$ = $WO_4^{2-}{}_{(aq)}$ + $CaCO_{3(s)}$ (calcite)	473-483 /12/
19)	$Fe^{2+}{}_{(aq)}$ + $2S^{2-}{}_{(aq)}$ = $FeS_{2(s)}$ (pyrite)	478-488 /11/
20)	$2Na^+{}_{(aq)}$ + $2Al(OH)_4^-{}_{(aq)}$ + $2H_2SiO_4^{2-}{}_{(aq)}$ = $4OH^-{}_{(aq)}$ + $3H_2O_{(l)}$ + $Na_2(AlSiO_4)_2 \cdot H_2O_{(s)}$ (desilication product, DSP)	478-493 /13/
21)	γ-$AlOOH_{(s)}$ + $OH^-{}_{(aq)}$ = $AlO_2^-{}_{(aq)}$ + $H_2O_{(l)}$	483-503 /14/
22)	α-$FeS_{(s)}$ + $2Cu^+{}_{(aq)}$ = $Cu_2S_{(s)}$ (chalcosite) + $Fe^{2+}{}_{(aq)}$	488-493 /7/
23)	α-$AlOOH_{(s)}$ (diaspore) + $OH^-{}_{(aq)}$ = $AlO_2^-{}_{(aq)}$ + $H2O_{(l)}$	488-503 /14/
24)	$CuFeS2(s)$ + $2Cu+ (aq)$ = $Cu2S(s)$ + $CuS(s)$ (covellite) + $Fe2+ (aq)$	493-503 /7/
25)	$Cu^{2+}{}_{(aq)}$ + $Fe^{2+}{}_{(aq)}$ + $2S^{2-}{}_{(aq)}$ = $CuFeS_{2(s)}$	498-513 /11/
26)	$NaAlSi_3O_{8(s)}$ (albite) + $8OH^-{}_{(aq)}$ = $Na_2(AlSiO_4)_2 \cdot H_2O_{(s)}$ + $4H_2SiO_4^{2-}{}_{(aq)}$	503-513 /15/
27)	$Cu^{2+}{}_{(aq)}$ + $2S^{2-}{}_{(aq)}$ + $Fe^{2+}{}_{(aq)}$ = $FeS_{2(s)}$ + $Cu_{(s)}$	503-518 /7/
28)	$FeS_{2(s)}$ + $2Cu^+{}_{(aq)}$ + $Cu_{(s)}$ = $Cu_2S_{(s)}$ + $CuS_{(s)}$ + $Fe^{2+}{}_{(aq)}$	508-523 /7/
29)	$FeS_{2(s)}$ + $2Cu^+{}_{(aq)}$ = $Cu_2S_{(s)}$ + $S_{(s)}$ + $Fe^{2+}{}_{(aq)}$	508-523 /7/
30)	$Zn^{2+}{}_{(aq)}$ + $S^{2-}{}_{(aq)}$ = $ZnS_{(s)}$ (sphalerite)	513-523 /11/
31)	$2NaAlSi_2O_6 \cdot H_2O_{(s)}$ + $Ca(OH)_{2(aq)}$ + $4H_2O_{\{l\}}$ + $3H_2SiO_4^{2-}{}_{(aq)}$ = $2NaOH_{(aq)}$ + $6OH-_{(aq)}$ + $CaO \cdot Al_2O_3 \cdot 7SiO_2 \cdot 6H_2O_{(s)}$ (CAPSPH)	518-528 /9/
32)	$FeWO_{4(s)}$(ferberite) + $CO_3^{2-}{}_{(aq)}$ = $WO_4^{2-}{}_{(aq)}$ + $FeCO_{3(s)}$(siderite)	518-533 /12/
33)	$NiO_{(s)}$ (bunsenite) + $2H^+{}_{(aq)}$ = $Ni^{2+}{}_{(aq)}$ + $H_2O_{(l)}$	518-528 /16/
34)	γ-$Al_2O_{3(s)}$ + $2OH^-{}_{(aq)}$ + $2H_2O_{(l)}$ = $Al_2O(OH)_6^{2-}{}_{(aq)}$	543-553 /17/
35)	α-$Al2O3(s)$ + $2OH^-{}_{(aq)}$ + $2H_2O_{(l)}$ = $Al_2O(OH)_6^{2-}{}_{(aq)}$	543-553 /14/
36)	$Na_2SO_4 \cdot 10H_2O_{(s)}$ (sodium sulfate decahydrate) = $Na_2SO_{4(s)}$ (sodium sulfate) + $10H_2O_{(l)}$	543-553 /18/
37)	$NaAlSiO_{4(s)}$ (nepheline) + $Ca(OH)_{2(aq)}$ + $OH^-{}_{(aq)}$ = $AlO_2^-{}_{(aq)}$ + $NaCaHSiO_{4(s)}$(sodium-calcium hydrosilicate, SCHS) + $H_2O_{(l)}$	553-563 /19/
38)	$CH_{4(g)}$ + $8Cl^-{}_{(g)}$ = $CCl_{4(l)}$ + $4HCl_{(g)}$	578-583 /20/
39)	$Na_2(AlSiO_4)_2 \cdot H_2O_{(s)}$ + $4FeOOH_{(s)}$ (goethite) + $6Ca(OH)_{2(s)}$ = $AlO_2^-{}_{(aq)}$ + $H_2O_{(l)}$ + $2Ca_3Fe_2(OH)_8(SiO_4)_{(s)}$ (calcium-ferrous hydrogarnet)	583-588 /21/
40)	$2C_2H_5OH_{(l)}$(ethanol) = $C_4H_{6(g)}$(1.3-butadiens)+ $2H_2O_{(g)}$ + $H_{2(g)}$	638-653 /22/

The tabular HTR equations are reduced to minimum number of the stoichiometric coefficients. As the temperatures are increased, the HTR numbering grows. The tabular data are impressive enough since they cover hydrometallurgical processes for production of nickel (reactions 3, 4, 19, 33), antimony (5), copper (6, 19, 22, 24, 25, 27-29), lithium (9, 12), manganese (16), lead (17), tungsten (18, 32), zinc (30), and aluminum, or rather, alumina (2, 7, 20,

21, 23, 26, 34-36, 39). In the last case, choosing formula of aluminate ions involved in the reactions is dictated by aluminate solutions concentration following the areas of aluminate ions domination in the $Na_2O-Al_2O_3-H_2O$ system a or manufacture of both astringents (1, 8, 10-15, 31) and chemicals (37, 38, 40). The digestion procedures dealt with the tabular equations are highly varied. There are among them hydration (1, 15), leaching (2, 5, 7, 9, 16, 18, 21, 23, 26, 32-36, 39), decomposition (6, 14, 24, 40), and dehydration (37). Hence, it can be said with reasonable confidence that the HTR tabulated are in the thorought character.

Finding the numerical values of K and T_1 of 6th Equation is the next stage for setting up a relationship between T_{suf} and $/\Delta_f S^0_m/$ of HTR. For this purpose, several references and handbooks have been used /24-32/. Absolute values of standard entropy change of the above HTR figured out with the Hess's law are shown in Table 2.4.2. The unknown parameters K and T_1 were computerized using the HTR temperature ranges of Table 2.4.1 and 6th Equation. As a result, the following formula has been derived

$$/\Delta_f S^0_m/ = 1.603 \, T_{suf} \cdot \ln(T_{suf}/405.21) \qquad \text{(7th Equation)}$$

where $1.603 = K = (\Delta C_p/298.15)$ and $405.21 = T_1$.

The temperature-entropy change relationship of HTR is diagrammed in Figure 2.4.2 in coordinates $/\Delta_f S^0_m/ - T_{suf}$.

Table 2.4.2. Thermodynamic characteristics of the hydrothermal reactions

No	$/\Delta_f S^0_m/$ J·mol^{-1}·K^{-1}	T_{suf}, K figured out by 7th Equation	Temperature range, K (from Table 4.2.1)
1	6.4	409	403-418
2	9.2	411	403-413
3	11.5	412	403-413
4	17.0	416	403-423
5	36.9	428	423-433
6	46.2	434	403-433
7	51.5	437	423-438
8	59.0	442	433-448
9	69.8	449	443-453
10	95.6	462	453-463
11	98.0	464	463-473
12	99.5	465	463-473
13	100.0	466	458-473
14	100.8	466	463-478
15	102.8	467	463-483
16	104.9	468	463-473
17	109.1	470	463-479
18	119.6	476	473-483
19	129.0	481	478-488
20	129.9	482	478-493
21	134.6	484	483-503
22	147.5	491	488-493
23	149.9	492	488-503
24	155.9	495	493-503
25	173.9	504	498-513
26	176.8	506	503-513
27	177.7	507	503-518
28	186.8	511	508-523
29	188.0	512	508-523
30	197.5	517	513-523
31	199.5	518	518-528
32	209.3	522	518-533
33	216.0	526	518-528
34	249.5	543	543-553
35	251.1	544	543-553
36	256.2	546	543-553
37	290.5	563	553-563
38	324.4	579	578-583
39	338.2	584	583-588
40	465.1	641	638-653

$/\Delta_f S^0_m/, J \cdot mol^{-1} \cdot K^{-1}$

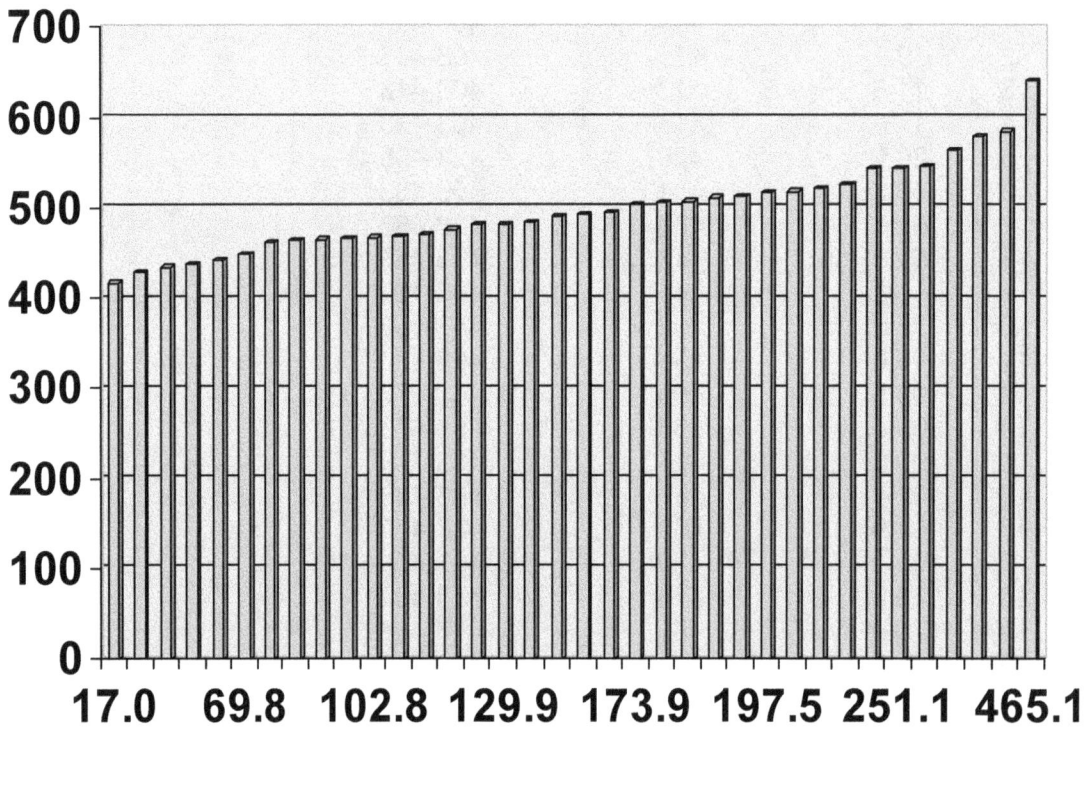

T_{suf}, K

Figure 2.4.2. Relationship between standard entropy change $/\Delta_f S^0_m/$ and sufficient temperature T_{suf} of HTR

2.4.3. Discussion. Corroboration of 7th Equation reliability may be gotten by the example of digestion of North Onega bauxite containing boehmite γ-AlOOH 34.5%, gibbsite $Al(OH)_3$ 14% and kaolinite $Al_2Si_2O_5(OH)_4$ 38%. Moreover, $Ca(OH)_2$ about 4% is added to the ore for speeding up the digestion. The gross equation of corresponding chemical reaction reduced to one mole of the initial summary aluminum hydroxide, with regard to the above minerals percentage, appears as follows:

-80-

0.763 γ-AlOOH(s) + 0.237 $Al(OH)_{3(s)}$ + 0.198 $Al_2Si_2O_5(OH)_{4(s)}$ + 0.356$Na^+_{(aq)}$ + 1.428$OH^-_{(aq)}$ + 0.06 $Ca(OH)_{2(s)}$ = $AlO_2^-_{(aq)}$ + 0.178$Na_2(AlSiO_4)_2 \cdot H_2O_{(s)}$ + 0.02$Ca_3Al_2(SiO_4)0.2(OH)_{11.2(s)}$ (calcium-aluminum hydrogarnet poorly saturated with silica*) + 0.036$H_2SiO_4^{2-}{}_{(aq)}$ + 1.581$H_2O_{(l)}$ (HTR #41)

This resulting reaction #41 involves contributions of each mineral digestion and, hence, it has the fractional stoichiometric coefficients.

    According to routine calculations the $/\Delta_f S^0_m/$ value for HTR #41 is equal to 167.9 $J \cdot mol^{-1} \cdot K^{-1}$ and, consequently, from 7th Equation, T_{suf}= 499.7K. This magnitude suits well to the temperature found practically /34/. The method of authors defined that the greatest possible alumina extraction yield has been achieved at temperature 498.15K. Bearing in mind an unevitable approximation of the experimental method of T_{suf} determination, both theoretically and bench-scale obtained results are, admittedly, in close agreement.

    Another example of 7th Equation use is concerned with hydrothermal alkaline leaching od DSP resulting the HTR #20 of aluminate solution desilication. The technique of DSP digestion should be similarly the alkaline leaching of nepheline (HTR #37). The reaction equation with regard to DSP is expected to be as follows:

$Na_2(AlSiO_4)_2 \cdot H_2O_{(s)}$ + 2$Ca(OH)_{2(s)}$ + 2$OH^-_{(aq)}$ = 2$AlO_2^-_{(aq)}$ + 2$NaCaHSiO_{4(s)}$ + 3$H_2O_{(l)}$ (HTR #42)

According to calculations, $/\Delta_f S^0_m/_{(42)}$=618.49 $J \cdot mol^{-1} \cdot K^{-1}$ and $T_{suf(42)}$= 702K. However, practically the digestion temperature for HTR #42 is evaluated 95-100K lower /19/. Once Fe-containing compounds would take part in the DSP digestion procedure, as in HTR #39, the temperature must be still less (584K) due to calcium-ferrous hydrogarnet formation. In the iron absence, it is impossible to carry on HTR #42 because of the following low-temperature (333-343K) nonhydrothermal interaction between DSP and slaked lime to produce calcium-aluminum hydrogarnet (CAHG):

$Na_2(AlSiO_4)_2 \cdot H_2O_{(s)}$ + 3$Ca(OH)_{2(s)}$ = $Ca_3Al_2(SiO_4)_2(OH)_{4(s)}$ (CAHG) + 2$NaOH_{(aq)}$ + $H_2O_{(l)}$

In its turn, CAHG enters upon digestion procedure into hydrothermal reaction with solution constituents giving SCHS as follows:

$Ca_3Al_2(SiO_4)_2(OH)_{4(s)}$ + 3$NaOH_{(aq)}$ + $H_2SiO_4^{2-}{}_{(aq)}$ = 2$AlO_2^-_{(aq)}$ + $NaCaHSiO_{4(s)}$ + 3$H_2O_{(l)}$ (HTR #43)

The design parameters are $/\Delta_f S^0_{m/(43)}$ = 385.6 $mol^{-1} \cdot K^{-1}$ and $T_{suf(43)}$ = 606K what proves experimental data and, consequently, confirms 7th Equation reliability.

---

*-$S_f^0{}_m$ value of this chemical is substance is taken from Ref. /33/

2.4.4.Conclusion. The regularity proposed makes a contribution to knowledge of chemical thermodynamic of the closed system. The possibility for determination of sufficient temperatures of hydrothermal technique permits to forecast the necessary changes of operating temperature under reactants composition variation. Moreover, it is reasonable using the approach found to solve the tasks as follows:

-To discover or make clear the chemistry of various HTR by the known T_{suf} values;

-To establish the temperature of depth natural HTR studying mineral composition of the rocks;

-To forecast the depth ores composition taking the deposit temperature;

-To calculate ore make more accurate the thermodynamic quantities of the HTR reactants;

-To develop the industrial processes involving HTR.

References

1.Kuznetsov,S.I.; Derevyankin,V.A. Physical Chemistry of the Process of Alumina Production by the Bayer Method. Metallurgiya: Moscow. 1964, 352 p.

2.Eryomin,N.I.; Naumchik,A.N.;Kazakov,V.G. Processes and Apparatus for Alumina Production. Metallurgiya: Moscow. 1980, 360 p.

3.Brodyansky,V.M.; Fratcher,V.; Mikhalek,K.; The Excergetic Method and Its Applications. Energoatomizdat: Moscow. 1988, 288 p.

4.Gorshkov,V.S.; Timashyov,V.V.;Savel'yev,V.G. Methods of Physicochemical Analisys of Astringents. Vyssh. Shkola: Moscow. 1981, 335 p.

5.Sutyrin,Yu.E.;Zverev,L.V.; Nitric Acidic Autoclave Decomposition of Products of High-Silicon Bauxites Benefication. Izv. Vyssh. Uchebn. Zaved., Tsvetn. Met. 1975, 2, p. 45-47.

6.Gudima,N.V.; Shein, Ya.P. Brief Handbook on Nonferrous Metallurgy. Metallurgiya: Moscow. 1975, 536 p.

7.Tseft, A.L. Hydrometallurgical Methods for Processing Polymetallic Raw Material. Nauka: Alma-Ata. 1976, 332 p.

8.Kolenkova,M.A.; Krein,O.E. Metallurgy of Less-Comon and Light Metals. Metallurgiya: Moscow. 1977, 360 p.

9.Barrer,R. Hydrothermal Chemistry of Zeolites. Academic Press: London. 1982, 424 p.

10.Maslenitsky,N.N.; Perlov,P.M.; Poprikailo,V.M.; Autoclave Processes of Nonferrous Metallurgy. Tsvetmetinformatsjya: Moscow. 1966, p. 200-220.

11. Rashidova, T.P.; Computerized Simulation of Processes of Hydrothermal Mineralogenesis. Master's Thesis: Irkutsk. 1982.

12. Maslenitsky, I.N.; Dolivo-Dobrovol'sky, V.V.; Dobrokhotov, G.N.; et al. Autoclave Procedures in Nonferrous Metallurgy. Metallurgiya: Moscow. 1969, 349 p.

13. Manvelyan, M.G.; Khanamirova, A.A. Desilication of Alkaline Aluminate Solutions. Academiya Nauk: Erevan. 1973, 300 p.

14. Pevzner, I.Z.; Rayzman, V.L. Autoclave Procedures in Alumina Production. Metallurgiya: Moscow. 1983, 128 p.

15. Manvelyan, M.G.; Nadzharyan, A.N.; Babayan, S.A.; Arevshatyan, M.S. Behavior of Main Minerals of Nepheline-Syenitic Rock in the Course of Its Chemical Beneficiation, Chemistry and Technology of Alumina. Sovnarkhoz: Erevan. 1964, p. 163-175.

16. Utkin, N.I. Nonferrous Metallurgy. Metallurgiya: Moscow. 1990, 448 p.

17. Ni, L.P.; Rayzman, V.L. Combining Methods of Processing Low-Grade Aluminum-Containing Raw Material. Nauka: Alma-Ata. 1988, 256 p.

18. Gorbanyov, A.I.; Nikolina, V.Ya. Sodium Sulfate. Goskhimizdat: Moscow. 1964, 234 p.

19. Sazhin, V.S. New Hydrochemical Methods of Complex Processing Aluminosilicates and High-Silicon Bauxites. Metallurgiya: Moscow. 1988, 213 p.

20. Petrov, A.A.; Bal'yan, Kh.V.; Troshchenko, A.T. Organic Chemistry. Vyssh. Shkola: Moscow. 1973, 623 p.

21. Ni, L.P.; Goldman, M.M.; Solenko, T.V. Processing High-Ferrous Baixites. Metallurgiya: Moscow. 1979, 248 p.

22. Mukhlenov, I.P.; Averbukh, A.Ya.; Kuznetsov, D.A., et al. General Chemical Technology, Part 2. Vyssh. Shkola: Moscow. 1977, 288 p.

23. Rayzman, V.L.; Vlasenko, Yu.K.; Pevzner, V.I. A Study of Boundaries Between Stability Fields of Different Aluminate Ions in the System Sodium Oxide-Alumina-Water, Kompleksn. Ispol'z. Miner. Syr'ya. 1986, 7, p. 61-65.

24. JANAF Thermochemical Tables. U.S. National Bureau of Standard Publication, NS RDS-NBS 37. 1970.

25. Thermodynamic Properties of Individual Substances: Editor Glushko, V.P., Volumes 1-4. Nauka: Moscow. 1979-1981.

26. Cox, J.D.; Washington, D.D.; Medvedev, V.A. CODATA Key Values for Thermodynamics. Hemisphere Publishing Corp.: New York. 1989.

27. Karapet'yants, M.Kh.; Karapet'yants, M.L. The Principal Thermodynamic Constants of

Both Inorganical and Organical Matters. Khimiya: Moscow. 1968, 417 p.

28.Ryabin,V.A.; Ostroumov,M.A.; Svit,T.F. Thermodynamic Properties of Substances. Khimiya: Leningrad. 1977, 392 p.

29.Bulakh,A.G.; Bulakh,K.G. Physicochemical Properties of Minerals and Constituents of Hydrothermal Solutions. Nedra: Leningrad. 1978, 167 p.

30.Rayzman,V.L. Determination of the Thermodynamic Characteristics of Aluminate Ions., Tsvetn. Met.. 1985, p. 60-62.

31.Kireev,V.A. Methods of Practical Calculations in Thermodynamics of Chemical Reactions. Khimiya: Moscow, 1975, 536 p.

32.Babushkin,V.I.; Matveev,G.M.; Mchedlov-Petrosyan,O.P. thermodynamics of Silicates: 4th Edition. Stroiizdat: Moscow. 1986, 408 p.

33.Alekseev,A.I. Calcium Hydroaluminates and Hydrogarnets. Len. Gos. Universitet: Leningrad. 1985. 64 p.

34.Leiteizen,M.G.; Bitner,A.A. Desilication and Digestion of North Onega Bauxite. Transactions: VAMI: Leningrad. 1970, 70, p. 120-125.

Chapter 3. INNOVATIONS IN PROCESSING LIQUID ALUMINATE INTERMEDIATES AND WASTES

3.1. Effect of Deep Aluminate Solutions Desilication on Alumina Yield in the Precipitation Procedure. Almost all of the alumina in the world is produced from bauxite by the Bayer process. This method, covered in Introduction and flow charted in Figure 1.1, has obvious advantages compared with other technologies (e.g., sintering or acid process) due to lower energy and material consumption [1]. It should be recalled that the SiO_2-containing minerals of bauxite react with an alkali-aluminate liquor to yield sodium hydroaluminosilicate (SHAS). This interaction causes loss of alumina, with the red mud separated from the solution by clarification. The reminder of the SiO_2 in solution is controlled by the equilibrium state of the Na_2O-Al_2O_3-SiO_2-H_2O system [2].

Silicon State and Behavior in Aluminate Solution. The quantities of equilibrium SiO_2 concentration in aluminate solution at a temperature of 90°C are shown in Table 3.1.1.

Table 3.1.1. Equilibrium composition of solution in the system
Na_2O-Al_2O_3-SiO_2-H_2O at 90°C [6]

Concentration (g/L)			Ratio		Associated stage of the Bayer process
Na_2O	Al_2O_3	SiO_2	$Na_2O:Al_2O_3$(mol)	Al_2O_3 SiO_2 (mass)	
249.0	102.4	0.7	4.0	146.3	Preparation
261.0	248.0	2.385	1.74	140.0	Digestion
109.2	108.5	0.675	1.66	163.0	Settling
125.0	51.4	0.3	4.0	171.3	Precipitation
354.4	332.0	5.4	1.74	61.5	Evaporation

The highest $Al_2O_3:SiO_2$ mass ratio conforming with the equilibrium state is equal to 173.1:1, while a conventional $Al_2O_3:SiO_2$ mass ratio of aluminate solution before the precipitation procedure lies between 250:1 and 300:1, resulting in the precipitation of aluminum hydroxide that is low enough in SiO_2 for smelting. In the modern view [2,3], the interaction between silicate and aluminate components of solution is described by the following reaction equations:

$[SiAl_nO_{2(n-1)}(OH)_m]^{(n+m)-}_{(aq)} + 2(n-1)H_2O_{(L)} =$

(P-ASHC)

$[SiAlO_4(OH)_m]^{(1+m)-}_{(aq)} + (n-1)Al(OH)_4^-_{(aq)}$ (1)

(M-ASHC)

$[SiAl_nO_{2(n-1)}(OH)_m]^{(n+m)-}_{(aq)} + 2Na^+_{(aq)} + (4n-3)H_2O_{(L)} =$

(P-ASHC)

$$2m(OH)^-_{(aq)} + Na_2(AlSiO_4)_2.H_2O_{(s)} + 2(n-1)Al(OH)_4^-{}_{(aq)} \qquad (2)$$

(SHAS)

where P-ASHC and M-ASHC = poly (P) and mono (M) - aluminum-silicon hydroxycomplex, respectively.

The first reaction takes precedence as the temperature is increased during the digestion and evaporation stages of the Bayer-process. The second reaction has an advantage during aluminate liquor cooling, before its clarification and precipitation. Consequently, the heat exchangers of evaporators and digesters are prone to fouling with SHAS-deposition [4]. On the other hand, the presence of (P&M)-aluminum-silicon hydroxycomplexes in aluminate solution during cooling causes the viscosity to increase and hence tends to slow down the settling of red mud and aluminum hydroxide. Therefore, more complete desilication of aluminate solution has to improve the above-mentioned stages. Furthemore, it is reasonable to suppose that the precipitation rate and Al_2O_3 yield are retarded because of too high SiO_2 content in an initial aluminate solution.

Impact of SiO_2 Concentration on the Precipitation of Alumina Hydroxide. It was detemined that the precipitation rate and Al_2O_3 yield depend directly on the behavior of the silicon dissolved in aluminate liquor [5]. As shown in Table 3.1.2, the precipitation of aluminum hydroxide from aluminate solution low in SiO_2 (Al_2O_3:SiO_2 mass ratio = 1,130:1) begins within the first three hours. After the Al_2O_3:SiO_2 mass ratio in aluminate solution is reduced to 18.7:1, the incubation period extends to over 12 hours. The SiO_2 is essentially not transferred from liquor to precipitate during this time. A further increase in time to 36 hours leads to a steep lowering concentration, by 0.45g SiO_2 /L less than at the beginning of precipitation. The SHAS is a resulting product of the aluminate solution desilication. It is important to note that the tests were carried out with the seed ratio reduced from the typical value 2-2.5:1 to 0.1:1 in order that to deliberately decrease the precipitation rate, trace the silicon behavior, without interference from additional concentration on Al_2O_3 yield. This assumption is supported by an empirical equation proposed by Leiteizen [3], which can be depicted as

$$A = 2000:/(A/S)_{mass}.(N/A)_{mol}/ \qquad (3)$$

where A, S, and N = concentrations of Al_2O_3, SiO_2, and Na_2O, respectively (g/L); 2000 = average constant value.

Based on Equation (3), the amount of Al_2O_3 in aluminate solution diminishes progressively as its parameters $(A/S)_{mass}$ and $(N/A)_{mol}$ are elevated. Whereas a correlation between A and $(N/A)_{mol}$ has received the most study, the function $A = f/A/S)_{mass}/$ needs to be examined closer. It is likely that the ratio $(A/S)_{mass}$ of aluminate liquor before the precipitation has to be raised from 300:1 to significantly greater above 1,000:1 and probably up to 3,000:1 or greater in order to achieve an increase of Al_2O_3 yield and a radical reduction in precipitation time. Technology for more complete desilication of the aluminate solution has been developed to fit the sintering process of alumina refining [2,3, 7-9].

Table 3.1.2.Alumina yield during the precipitation procedure with the seed ratio 0.1:1 by the Al_2O_3:SiO_2 mass ratio in an initial aluminate solution 1130:1 (1) and 18.7:1 (2)

The length of test (hrs)		Concentration (g/L)		Alumina yield (%)
		Na_2O	Al_2O_3	
(1)	0	130.5	130.9	0
	3	132.3	117.0	12.0
	12	135.5	81.5	40.2
	24	134.8	71.2	47.4
	36	134.6	65.5	53.1
	48	134.5	59.8	55.8
(2)	0	123.6	121.0	0
	3	120.4	119.1	0
	12	123.2	119.2	0
	24	125.2	70.5	39.7
	48	125.1	52.3	52.6

Desilication of Aluminate Solution Produced by Leaching of Sintered Material. A simplified flow chart for the sintered method is shown in Figure 3.1.1.

Raw material high in silica is mixed, during its processing, with limestone and recirculated soda solution and the resulting blend is sintered into solid product containing sodium aluminate and bicalcium silicate. The sintered material is leached with process water and the diluted aluminate is separated from the slurry. Since aluminum hydroxide precipitation has to be carried out by total carbonization of aluminate solution, it should be thoroughly purified from the silicon that possesses the ability to co-precipitate in association with aluminum in SHAS form. Therefore, desilication is performed in two stages. The $(A/S)_{mass}$ of the solution obtained as a result of the 1st stage is similar to that of the aluminate liquor in the Bayer process. The second stage of desilication is carried out with a lime addition. The procedure ends with aluminum-calcium hydrogarnet (ACHG) formation and is described by the following reaction equation:

$$x[SiAl_nO_{2(n+1)}(OH)_m]^{(n+m)-}_{(aq)} + 3CaO_{(s)} + (3 - 2x + 2nx)H_2O_{(L)} =$$

$$Ca_3Al_2(SiO_4)_x(OH)_{(12-4x)(s)} + (nx-2)Al(OH)_4^-_{(aq)} + (mx+2)OH^-_{(aq)}$$

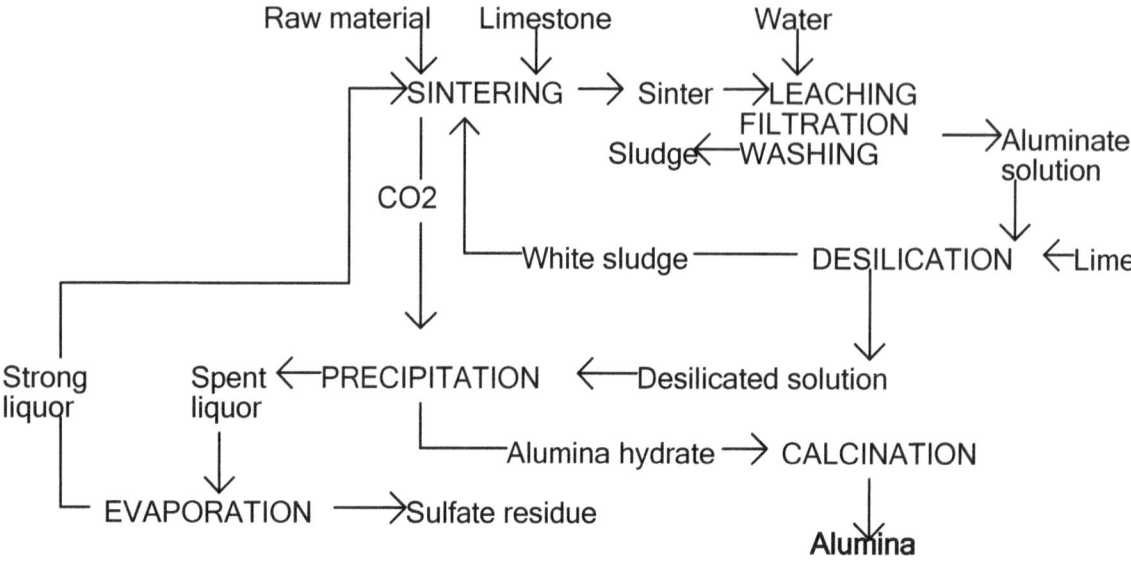

Figure 3.1.1. Flow chart for processing high-silicon raw material by the sintering method

The second desilication stage requires between 1.5 to 2 hours at 90-95°C. Lime is added to the initial aluminate solution at a rate of 16 to 18 g per liter. The resulting $(A/S)_{mass}$ ratio of the solution varies routinely between 1,000:1 and 1,200:1. This value provides an acceptable quality of the common commodity alumina. However, the more complete silicon precipitation should be conducted for improving the Bayer process. The preconditions exist for solving this problem. Table 3.1.3 shows the influence of calcium-containing reagents and their chemical composition on desilication. An $(A/S)_{mass}$ ratio of 2,500:1 to 6,000:1 has been reached by the addition of ground limestone and hydrocarbocalcium aluminate (HCCA) $Ca_4Al_2O_6.CO_3.11H_2O$ [10] to the silicate-containing aluminate solution at 90°C. The technology of more complete desilication of aluminate solution and preparation of the HCCA was tested at the VAMI pilot plant (St. Petersburg, Russia) and some of Russia's alumina refineries that process nepheline into alumina and byproducts.

Table 3.1.3. The ending Al_2O_3:SiO_2 mass ratio of the desilicated aluminate solution, adjusted to 1, as a function of composition and consumption of calcium-containing precipitant (a duration of stirring ranged from 3 to 6 hours)

Precipitant dosage recalculated to active CaO (g/L)	Precipitant name and formula or symbol			
	Lime $Ca(OH)_2$	Lime+HCCA	Limestone $CaCO_3$	HCCA
1.08	450	450		600
4.0	450	500	600	950
8.0	470	800		2400
9.0			2500	
12.0	530	1100		400
16.0	1000	1700		6000

Feasibility of Applying Calcium-Containing Precipitants for More Complete Desilication of Aluminate Solutions in the Bayer Process. Input of calcium carbonates, such as limestone or HCCA, in the Bayer process causes, at first glance, serious complications related to caustic conversion into soda. However, the aluminum-calcium hydrogarnet (ACHG), that is the end product of the 2nd desilication stage, has the ability to caustize soda. The process proceeds according the reaction [11]:

$$Ca_3Al_2(SiO_4)_x(OH)_{(12-4x)(s)} + (3x - 3y -2) Na^+_{(aq)} + 3CO_3^{2-}_{(aq)} + [0.5(3x - y) - 2] H_2O_{(L)}$$
$$= 3CaCO_{3(s)} + 0.5(x - y) Na_2[AlSiO_4]_2.H_2O_{(s)} + (2 - x - y)Al(OH)_4^-_{(aq)} + yH_2SiO_4^{2-}_{(aq)}$$
$$+ (4 - 2x)OH^-_{(aq)}$$

Alumina yield from ACHG-containing sludge of the 2nd desilication stage reaches the range of 75-89% by treating sludge for 60 minutes with the soda solution at 90-100°C, Na_2CO_3 concentration from 170 to 275 g/L, and liquid:solid ratio between 7.5:1 and 10:1. The resulting materials are characterized by the following composition [3]:

1) Restored solution Al_2O_3 21 g/L, NaOH 40 g/L;

2) Carbonate calcium sludge: SiO_2 0.6%, Al_2O_3 1.7%, CaO 57%, CO_2 30%, Na_2O 0.8%, and L.O.I. 38%. It is of considerable industrial interest to apply other precipitants, such as non-carbonate tricalcium hydroaluminate $Ca_3Al_2(OH)_{12}$ and magnesium-containing materials for more complete desilication of solution in the Bayer process.

Conclusion. Improving the parameters of the Bayer process by more complete desilication of aluminate liquor will require more complex technology for adding lime or limestone to bauxite or countercurrent to one another for preventing the overconsumption of lime. In spite of this, the possibility of reducing heat exchanger fouling and precipitation time by the additional desilication of aluminate solution makes this method the most promising for alumina refining development.

References:

1.Academic American Encyclopedia, Danbury (CT), Groiler, Inc., 1992, 1, p. 316-317.

2.I.Z. Pevzner and A.N. Makarov, "Desilication of Aluminate Solutions," Moscow, Metallurgiya, , 1973, 112 p.

3.M.G. Manvelyan and A.A. Khanamirova, "Desilication of Alkaline Aluminate Solutions," Erevan, Acad. Nauk of Arm SSR, 1973, 300 p.

4.H. Muller-Steinhagen, M. Jamialahmaadi, and B. Robson, "Understanding and Mitigating Heat Exchanger Fouling in Bauxite Refineries," Journal of Metal, 1994, 11, p. 36-41.

5.L.G. Romanov, L.P. Ni, and E.F. Osipova, "About the Mechanism of Silica Impact on Aluminate Solutions Decopmosition," Innovations to Technology for Alumina Precipitation from Aluminate Solutions: Kazan' University, 1975, p. 13-22.

6.L.P. Ni and V.L. Rayzman, "Combined Methods for Processing Low-Grade Aluminum-Containing Raw Material," Alma-Ata, Nauka, 1988, 256 p.

7.V.L. Rayzman, Yu.K. Vlasenko, L.S. Nisse, et al, "Sodium Aluminate and Hydroaluminate - Manufacture and Use," Rostov, University, 1991, 120 p.

8.V.M. Sizyakov, V.I. Korneev, and V.V. Andreev, "Improvement of Alumina and Byproducts Quality During Nepheline Processing," Moscow, Metallurgiya, 1986, 118 p.

9.V.L. Rayzman, "Thermodynamic Analysis of Interaction of Calcium Hydrogarnet and NaOH Solution," Journal of Applied Chemistry of the USSR, 1986, 59(1), p. 234.

10.Ja.B. Rosen, V.A. Volkov, V.L. Rayzman, et al, "Desilication of Wastewater of Leningrad VAMI Pilot Plant," Nonferrous Metallurgy (Tsvetmetinformatsiya), 1980, 10, p. 32-33.

11.I.Z. Pevzner and V.L. Rayzman, "Autoclave Processes in Alumina Refining," Moscow, Metallurgiya, 1983, 128 p.

3.2. Extracting Sodium Aluminate from Aluminate Solitions and Liquid Wastes.

Sodium aluminate (SA) formulated as $NaAlO_2$, $Na_2O \cdot Al_2O_3$, or $Na_2Al_2O_4$ [1302-42-7] is the most widespread alkali metal aluminate. Other compounds of this group like potassium or lithium aluminates have significantly minor commercial use. SA is a constituent of solution formed in the course of the Bayer process, which involves the alkaline extraction of aluminum from bauxite into an aluminate solution form, followed by the precipitation of aluminum hydroxide and its calcination into alumina [1]. Aluminate solution can also be obtained by leaching of the sintered material containing the solid SA and an insoluble dicalcium silicate [2], as well as by alkaline etching of an aluminum surface. To produce SA as a commodity, aluminum hydroxide or metallic aluminum is dissolved in liquid caustic. The obtained aluminate solution is treated by drying, crystallization, or stabilization.

Sodium Aluminate Use. SA is, mainly, in demand for use in effective subacid water purification [1]. Producers of paper, paint pigments, alumina-containing catalysts, dishwasher detergents, ingot molds, molecular sieves, concrete and others are among SA users [3]. Main fields of the SA application are shown in Table 3.2.1.In addition, if drinking water is treated with SA and aluminum sulfate, the formed flocs are a good absorbent based on the surface-active aluminum hydroxide, making it possible to remove dye-stuffs and colloids.

Both reagents are added as an aqueous solution containing 5-10% SA and 2.5-5% aluminum sulfate [1,3,4]. Moreover, SA is used for purifying water used in swimming pools; potassium aluminate can also be used for water treatment under certain conditions (primarily an acceptable price)[12]. Apart from plug bore holes and waterproof concrete production, SA may be used for various hydraulic structures in order to seal and prevent cracks. In all applications, SA decreases corrosion in the apparatus and pipes. Both the formation and decomposition of SA solutions are defined by phase conversions in the sodium oxide-alumina-water system at various temperatures.

Table 3.2.1. Some examples of sodium aluminate application

Object	Procedure	Benefits	Reference
Fresh water purification	SA may be used both singly or in combination with other chemicals	Full precipitation of iron, manganese and organic matters has been achieved	1, 3, 4,5
Drilling mud prepa-ration for plug boreholes	SA is mixed with water glass and the mixture obtained is added to the drilling mud	Ground is hardened and sealed against possible water lea-kage	3, 4, 5
Production of waterproof concrete	SA is a constituent of many auxiliary matters added to concrete raw material	Concrete setting is sped up, the rigidity while watering the fresh concrete mostar is increased and concrete cold-resistance is improved	1
One-time ground silication	Mixture of sodium silicate and SA is injected into ground through well drilling	Increasing strength and decreasing compressibility of grounds	6
Paper and cardboard production	1. SA is built into process controlling the medium pH to provide the resin sizing precipitation 2. SA is applied to pulp processing	1. The resulting product is sized better and has higher than usual tensile strength. 2. Foaming of suspensions is reduced; swelling ability of the cellulose fiber is improved; fine dispersed fillers like titanium dioxide are fixed by paper more completely	7
Municipal and industrial - waste-water disposal	SA is added in li-quid or solid form	Removal of phosphates and water clarification, especially useful in wastewater low in alkali	1, 8, 9
Alumino - silicate [1327-36-2] preparation	Impregnating silica gel with SA and aluminum sulfate addition	There are porous crystalline aluminosilicates formed which are useful as adsorbents and catalysts support materials, i.e. molecular sieves	10, 11

Equilibrium and Transitions in the Na₂O-Al₂O₃-H₂O System. Many isotherms of this system are constructed in coordinates of Al_2O_3 (vertical axis) and Na_2O (horizontal axis) concentrations, based on evidence derived from experiments [13-17]. The presence of both left and right crossing curves (Figure 3.2.1) is apparent in the phase diagram.

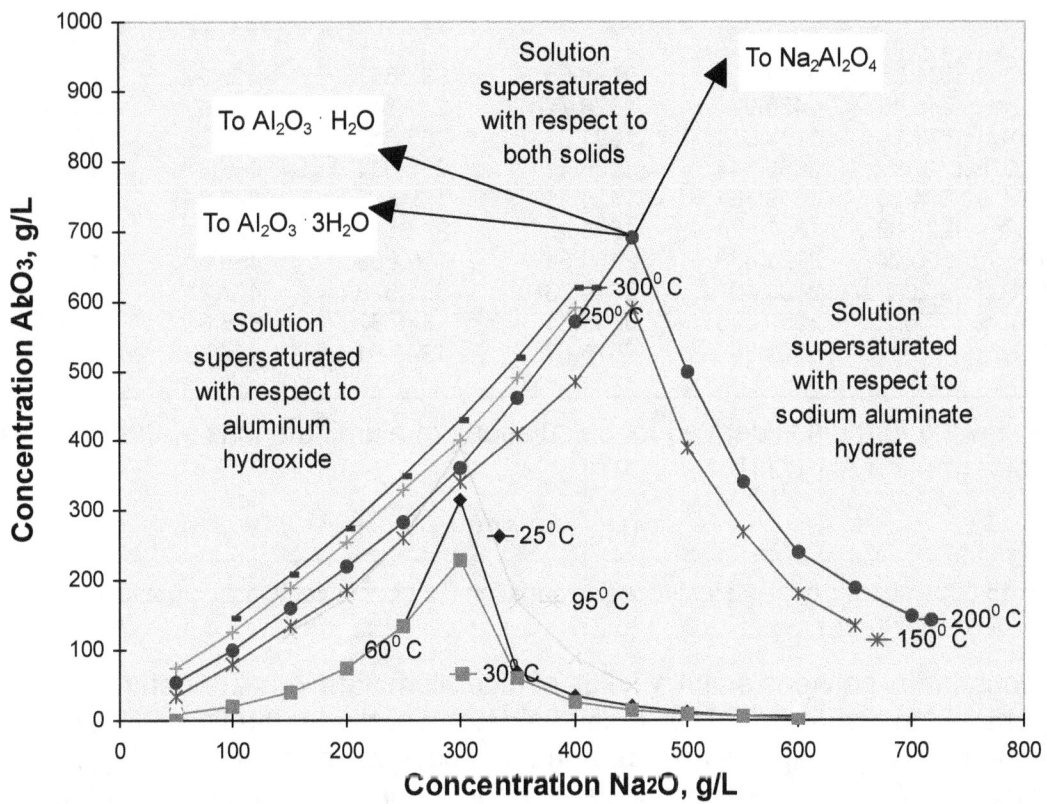

Figure 3.2.1. The phase diagram for the Na_2O-Al_2O_3-H_2O system

The higher the system temperature, the steeper are the curves branches and the higher is the Al_2O_3 equilibrium concentration. The aluminate solution stability field occurs within both left and right branches. The left portion of each family of curves is a line along which the solution is in equilibrium with aluminum hydroxide. The latter may be $Al_2O_3.3H_2O$ or $Al(OH)_3$ as gibbsite at temperatures up to 150°C and $Al_2O_3.H_2O$ or $AlOOH$ as boehmite at 150-250°C or diaspore at temperatures higher than 250°C. The right branches give conditions at which aluminate solution and sodium aluminate hydrate (SHA) formulated as $Na_2Al_2O_4.nH_2O$ (where n=2.5-3) are in equilibrium. The isotherms peak triple point [18] is the only point at which all three phases (i.e., solution, alumina hydrate and SHA) are in equilibrium. It has been suggested that the triple point of the system reached a maximum ionic strength (*I*) of aluminate solution [19]. At the present time, it is commonly recognized that aluminate solutions have ionic constitution [1,16]. The major standard thermodynamic quantities calculated for 10 aluminate anions are shown in Table 3.2.2 [20].

Table 3.2.2. Thermodynamic molar quantities of aluminate anions [20].

Ion	Entropy S°_f (J/K)	Enthalpy change $-\Delta H^\circ_f$(kJ)	Gibbs energy change $-\Delta G^\circ_f$ (kJ)	Heat capacity C_{p298} (J/K)	Coefficients of Equation $C_p = a + bT \cdot 10^{-3}$	
					a(J/K)	b(J/K)
AlO_2^-	104.6	911.05	853.966	89.365	-139.22	0.767
$Al(OH)_4^-$	245.46	1488.92	1328.774	241.345	12.76	0.767
$AlO(OH)_2^-$	175.1	1173.974	1091.3	165.355	-63.23	0.767
$Al(OH)_4O2H_2O^-$	385.1	2055.644	1803.57	367.091	96.08	0.909
$Al(OH)_5^{2-}$	234.36	1649.564	1421.96	1.345	227.24	0.767
$Al(OH)_6^{3-}$	223.62	1815.524	1515.15	27.654	229.8	-0.678
$Al_2(OH)_7^-$	500.94	2721.438	2484.54	796.858	459.65	1.131
$Al_2O(OH)_6^{2-}$	420.2	2601.48	2340.33	581.528	407.11	0.585
$Al_3(OH)_{10}^-$	756.78	3959.423	3640.31	1340.82	876.3	1.558
$Al_6(OH)_{24}^{6-}$	1470.6	8502.88	7573.57	1945.815	1379.33	1.9

The following formula is derived for calculations of aluminate ions standard molar heat of formation ΔH°_f [21]:

$$\Delta H^\circ_f = 15.609 \, M_i$$

where 15.609 - specific heat of aluminate ion formation (kJ per conventional unit) and M_i is the molecular weight of ion.

The boundaries between stability fields of nine aluminate ions are determined as diagrammed in Figure 3.2.2 [22]. It can be determined from the figure that only three single-charged anions may exist in the most diluted solutions: AlO_2^-, $Al(OH)_4^-$, and $AlO(OH)_2^-$. Other species are formed in the more concentrated medium. The calculations performed were based on the electrostatic theory of electrolyte solutions and conception ionic strength. The ionic strength was calculated using an aluminate solutions with Na_2O concentrations within 1-40% and molar ratio of Na_2O/Al_2O_3 ($\alpha_{caust.}$) of 1-18:1. The thermodynamic properties of sodium aluminate are in Reference 1. The Gibbs energy change of SHA formulated as $Na_2Al_2O_4.2.5 \, H_2O$ at 298.15K is 2890.9 kJ/mol [23]. Samples of SHA and SA were studied by ^{1}H, [27]Al and [23]Na nuclear magnetic resonance spectroscopy [24].

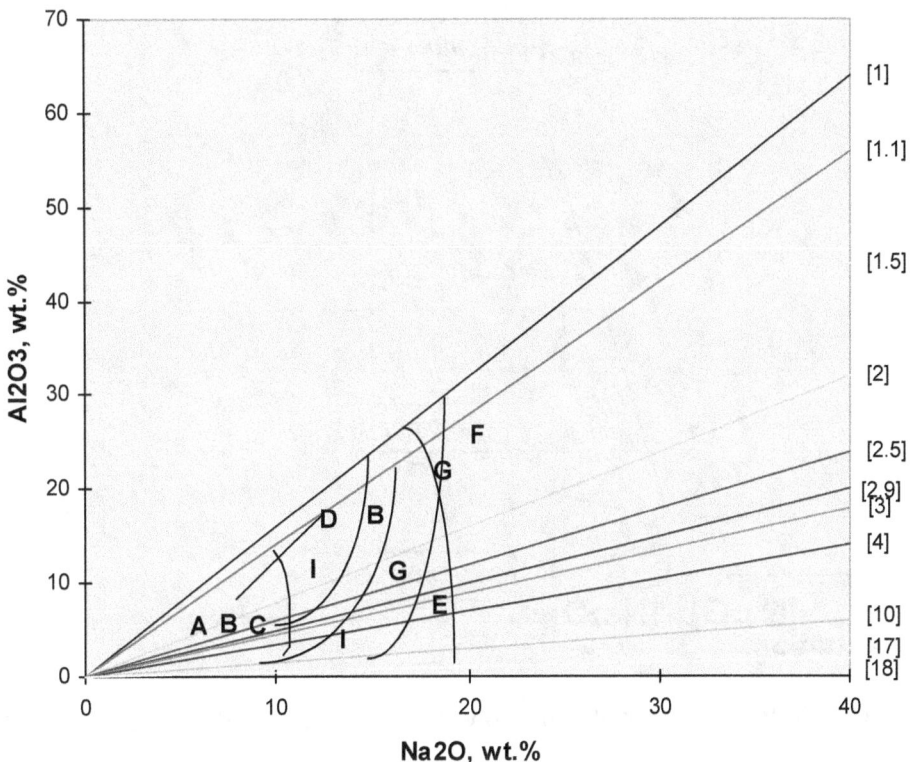

Figure 3.2.2. The bottom boundaries of aluminate ions stability in the Na_2O - Al_2O_3 - H_2O system [22].
where: A = AlO_2^-, B = $Al(OH)_4^-$, C = $AlO(OH)_2^-$, D = $Al(OH)_5^{2-}$, E = $Al(OH)_6^{3-}$, F = $Al_2(OH)_7^-$,
        G = $Al_2O(OH)_6^{2-}$, H = $Al_3(OH)_{10}^-$, I = $Al_6(OH)_{24}^{6-}$.
Numbers in parentheses[ ] are molar ratios of Na_2O : Al_2O_3 ($a_{caust.}$).

The proton-containing groups in the tested matters are hydroxide-ion $(OH)^-$. In
SHA samples, $Na_2(HO)_3AlOAl(OH)_3$ /shortly, $Na_2Al_2(OH)_6$/ predominates. The
basic structure of the SA sample is a face-centered cubic with a=0.568 nm; its
structure is assigned to the cubic space group Fd3m with Z=8 for the compound
$Na_2Al(OH)_5$. The precipitation of SHA from saturated aluminate solutions follows
the general regulations of crystallization. The procedure course is depicted in
Figure 3.2.3. The solution with the most suitable composition at point A is heated
and evaporated to the point B composition where the solution becomes
supersaturated with respect to the 60°C isotherm. During the temperature
reduction, the solution composition is changed on a straight line passing through
points F and B to the intersection at point C, which is the temperature designated
for SHA crystallization. Point F conforms to SHA composition.

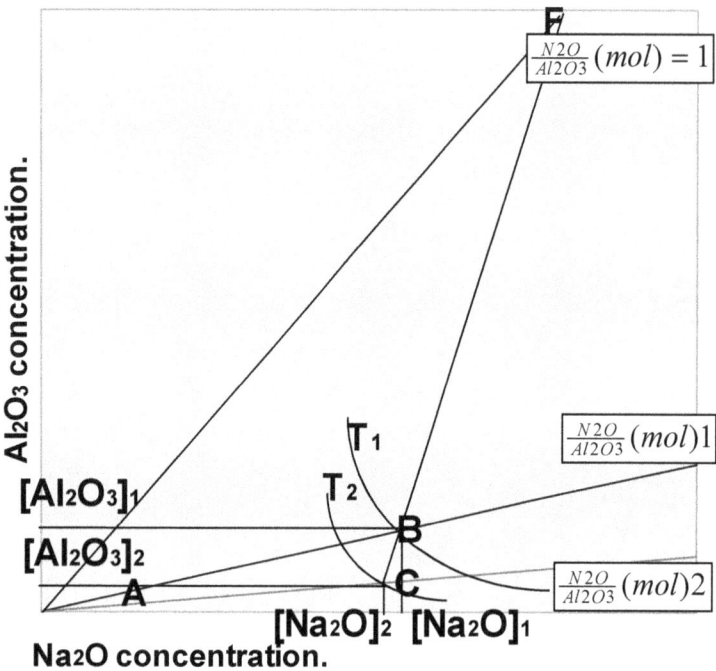

Figure 3.2.3. SHA crystallization in the framework of Na_2O - Al_2O_3 - H_2O system

The yield of anhydrous SA (wt %) is calculated by the formula:

SA yield = $100 \times ([Na_2O]_C \times [Al_2O_3]_B - [Na_2O]_B \times [Al_2O_3]_C) : ([Al_2O_3]_C \times [Na_2O]_C - (62:102)[Al_2O_3]_C)$ (1)

where the brackets represent the oxides concentration; B and C on the phase diagram indicate composition points; and numerals 62 and 102 are molecular weights for Na_2O and Al_2O_3, respectively.

In addition to isotherms, the following equation was set up to find any isotherm point as a function of solution composition and temperature [25]:

$$\log_{10}\alpha_{caust.} = A \cdot \log_{10}[Na_2O] - B \qquad (2)$$

where [Na2O] is the sodium oxide concentration, A and B are the temperature-dependent coefficients.

For example, at 50°C, coefficient A=5.55 and coefficient B=6.85. At 85°C, coefficient A=4.72 and coefficient B=6.18; at 135°C, coefficient A=4.25 and coefficient B=5.70 [25]. Interactions between the above-mentioned phases form a theoretical basis for manufacturing SA and SHA. Of special interest are the alumina-bearing intermediates and wastes as low-cost resources for SA production.

Processing Industrial Intermediate Materials into Sodium Aluminate. Sodium aluminate can be produced from the alumina-refinery intermediates, such as aluminate solutions and sintered materials. Aluminate solutions are obtained by the digestion of bauxite in the Bayer process or aluminosilicates in the hydrothermal method or by leaching SA-containing solid intermediate material formed in the sintering process. Both evaporation-crystallization and/or drying procedures may be used.

SHA Precipitation from Aluminate Solutions in the Bayer Process. The main object of this procedure is process improvement. From this standpoint, SHA is considered a byproduct. According to one of methods, spent liquor depleted by alumina hydrate precipitation is evaporated at 135°C to an NaOH concentration of 580-680 g/l. The evaporated solution is cooled to 95°C at 10 degrees/min., and SHA falling-out crystals are separated [26]. In this way, 70% of the dissolved sodium aluminate was recovered into the residue. This method was modified to precipitate SHA from aluminate solutions jointly with organic and salt matters [27]. The material balance of the process is represented in Figure 3.2.4.

Spent liquor was evaporated to an NaOH concentration of 385-400 g/l, then cooled to 85°C and held at this temperature for two hours to obtain soda hydrate precipitation. The formed residue was settled and the solution clarified was mixed with alumina hydrate. The prepared slurry was heated and evaporated to an NaOH concentration of 500-555 g/l simultaneously with dissolving of $Al(OH)_3$. The obtained strong aluminate solution was cooled to 85-95°C, subjected to four hours stirring, and precipitated SHA-containing crystals were filtered.

The summary impurities yield from initial spent solution were soda 90.2%; sodium chloride 50.7%; and organic matter 39.1%.

SHA Precipitation as an Application to the Sintering Process. The procedure is directed at separation of both sodium and potassium alkalies that are contained in the feed alkaline aluminosilicates (e.g., nepheline $Na_xK_{(1-x)} AlSiO_4$). [28] During sintering, the following interaction between nepheline and limestone takes place
$$Na_xK_{(1-x)} AlSiO_{4(s)} + 2CaCO_{3(s)} = xNaAlO_{2(s)} + (1-x)KAlO_{2(s)} + Ca_2SiO_{4(s)} + 2CO_{2(g)}$$
Under leaching by process water, Na and K alkali-aluminate solutions are formed. The caustic module ($\alpha_{caust.}$) of the solution that evaporated before its cooling must be no less than 3.5, otherwise, the formed aluminate suspension becomes too tough up to monolithic state [29].Therefore, the initial solution should be subjected to preliminary $Al(OH)_3$ precipitation followed by alkalization so as to provide α_{caust} = 4. The resulting solution is evaporated to an (Na+K)OH concentration of 713 g/l (calculated to NaOH), then cooled to 60°C, stirred at this temperature for six hours and filtered. A filtrate containing 670g (Na+K)OH/l, 25 g Al_2O_3/L, and the resulting uncombined SHA residue has α_{caust} equal to 1.68.

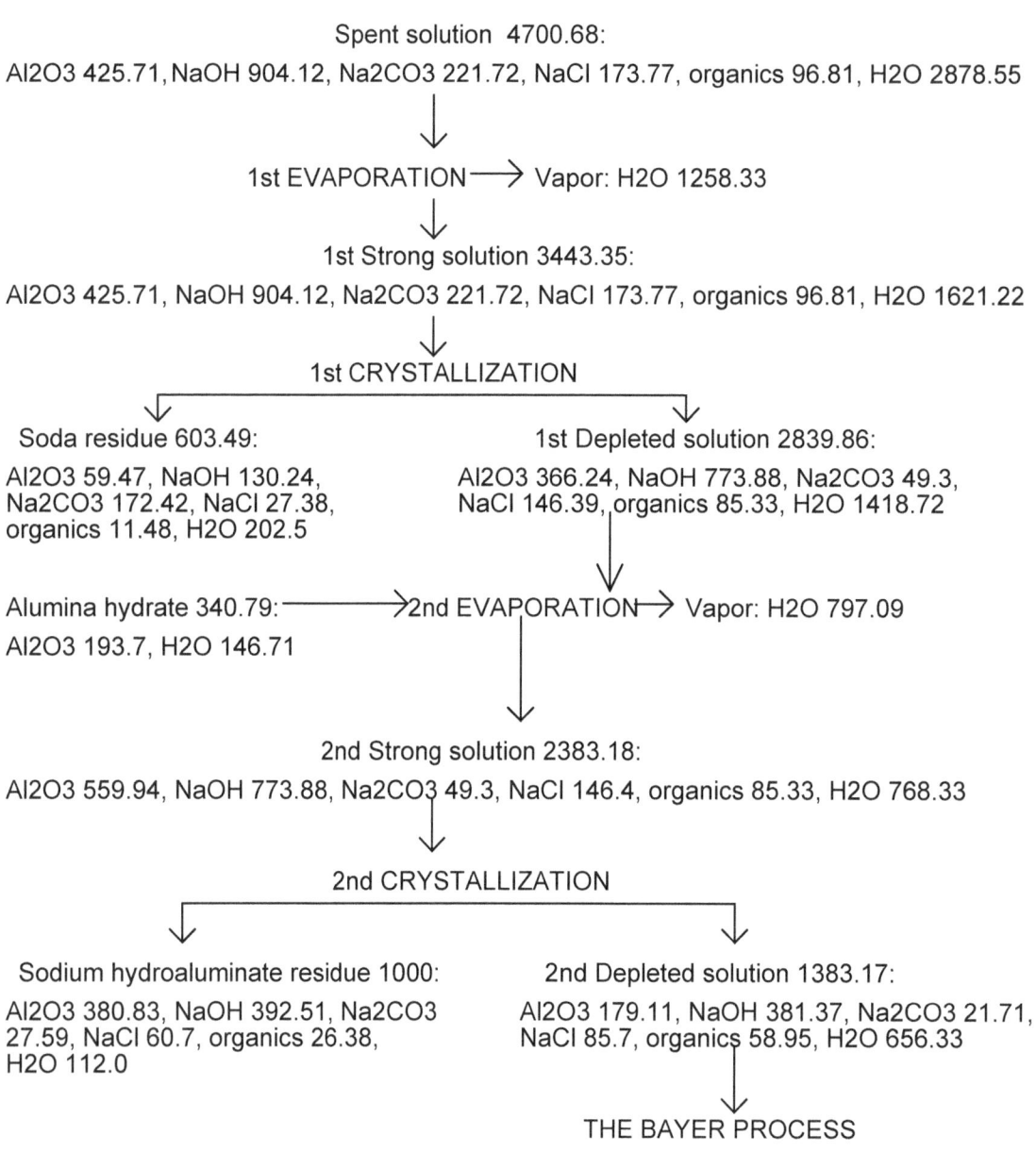

Figure 3.2.4. Material balance of process for purification of the Bayer process aluminate solutions from sodium carbonates, chlorides and organics through co-precipitation with SHA

The composition of the crystallized solids depends on the sodium and potassium contents in the evaporated solution. SHA is crystallized, only, if KOH concentration is no more than 60 mol. % of the NaOH and KOH total content in solution. With increasing KOH content to 60-75 mol. %, the mixture of SHA and PHA is formed. The compound $KAlO_2 \cdot 1.5 \, H_2O$ is crystallized only in the case that the KOH concentration would be more than 75 mol. %. Contrary to this is the minimum value of 3.5 α_{caust} for evaporated aluminate solution. A value parameter that is two times lower has been obtained experimentally [30]. This advantage is achieved by the following procedures. An initial aluminate solution containing (Na+K)OH 178-183 g/L (calculated to NaOH) and Al_2O_3 130-139 g/l (α_{caust} =1.68-1.75) is evaporated to 502-509 g (Na+K)OH/L, then cooled; alkali-metals hydroaluminate is precipitated. The optimal chilling temperature is equal to 60°C. The majority of potassium alkali remains in the depleted solution at this temperature. Moreover, the possibility exists for lowering the NaOH concentration in evaporated solutions from 710-715 g/L, as Reference [29] recommends, to 500-510 g/l, which will precipitate the alkaline-metals hydroaluminate [31]. The SHA crystallization is a first-order reaction governed by the Arrenius equation in the range of α_{caust} from 1.7 to 2 and at temperatures between 60°C and 90°C [32]. The activation energy value of SHA crystallization is calculated to be equal to 26 kJ/mol. [33].

SHA Precipitation with Reference to the Hydrothermal Process. Due to this procedure, a conversion of high-module aluminate solutions to low-module in α_{caust} liquors is carried out [34-36]. Aluminum can be extracted from aluminosilicate into solution under certain conditions, one condition being that the minimum α_{caust} value of obtained solution equal 10-11. In accordance with the phase diagram of the sodium oxide-alumina-water system (Figure 3.2.1), it is impossible to precipitate $Al(OH)_3$ from such high-module solutions by the conventional hydrolysis used in the Bayer technology. Therefore, a two-stage conversion has been conducted for the decomposition of alkali-aluminate solutions [34]. In the first stage, SHA having low α_{caust} = 1.5-1.8 is precipitated by solution evaporation to an NaOH concentration greater than 650 g/L, with subsequent cooling to 45-60°C and the obtained residue separation from the depleted solution. The resulting solids are dissolved in process water to become the low-module liquor that is good for alumina hydrate precipitation. It has been subjected that product precipitation will speed up with seeding evaporated high-module solutions with SHA as $Na_2Al_2O_4 \cdot 2.5 H_2O$ crystals [37]. The solids are separated from the mother liquor and washed with a solution having NaOH concentration lower than that in the mother liquor and an Al_2O_3 content of 60 g/L. Instead of in-depth evaporation of the alkaline solutions, the alkali extraction with methanol was tested on a bench-scale [38]. Two resulting products were formed. The top liquor is a mixture of caustic solution with methanol and the bottom residue is SHA of α_{caust} =1-1.2. The method of alcoholic alkali extraction with reference to hydrothermal technology is developed [39] and tested on the pilot-plant scale [40]. Due to the mastered process, α_{caust} of the aluminate solution was reduced from 11 to 1.6, matching the Bayer process conditions. The lowest aliphatic alcohols were tested as extractants to precipitate not only SHA and PHA, but also lithium hydroaluminate [41]. The obtained data were used for utilization of various alumina-bearing wastes.

Sodium Aluminate Recovery from Industrial Liquid Wastes. Wastewater of alumina refinery may be a practically inexhaustible source for manufacturing SA. For this purpose,

91

a processing plant was installed and put into operation [3]. Wastewater consisting of 20-25 g/L NaOH is pre-evaporated to the following concentrations: 117.6 g/L NaOH, 2 g/L Na_2CO_3, 12.4 g/L KOH, 65.2 g/L Al_2O_3, 0.45 g/L SiO_2, and 6.2 g/L SO_3. The solution is desilicated for six hours at 90°C to a mass ratio of alumina to silica 2500:1 [42]. Treatment of the obtained solution is carried out in a unit schematically depicted in Figure 3.2.5.

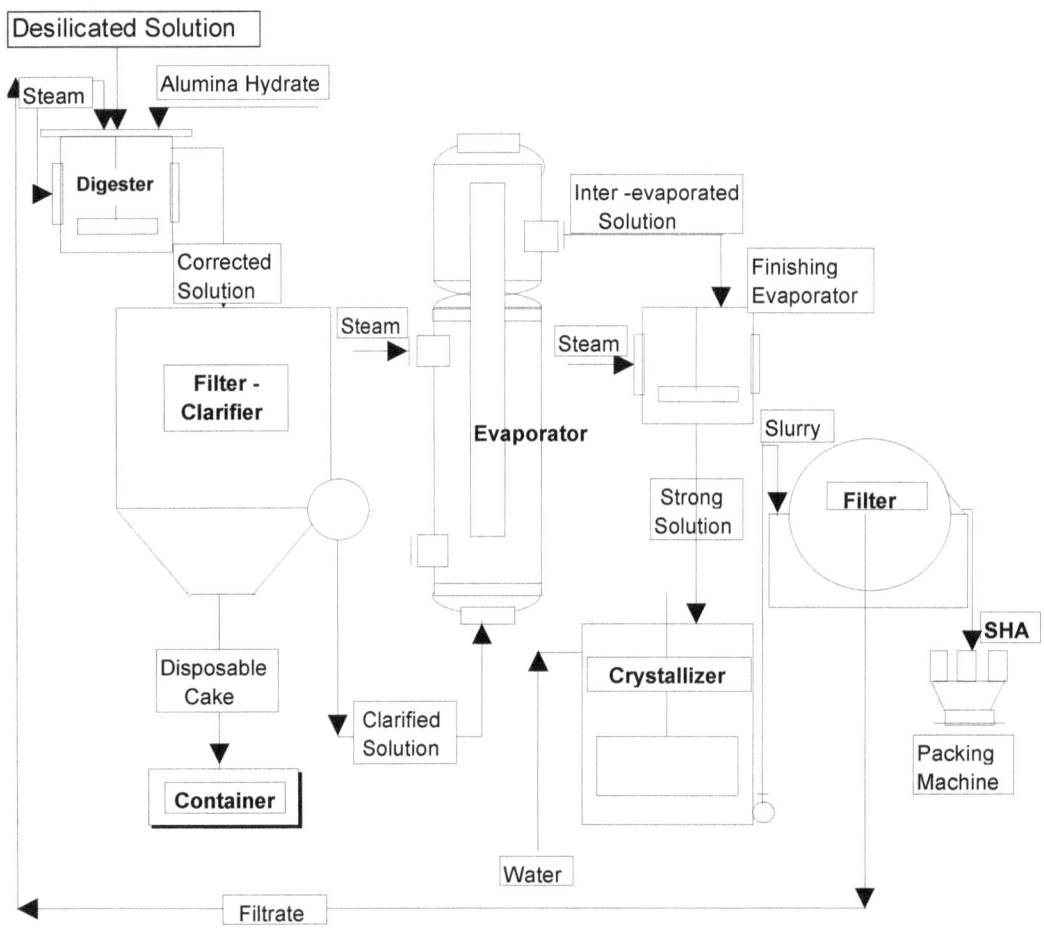

Figure 3.2.5. Flow chart of processing plant for utilization of alumina-bearing wastewater to sodium hydroaluminate (SHA)

At first, the solution is corrected to lower α_{caust} = 1.7-2 (instead of 2.4-2.5) by alumina hydrate digestion. The corrected liquor is clarified by filtration and evaporated in two stages. The first stage is conducted on a quadruple-effect evaporator to reach NaOH concentration of 320-390g/L and precipitate impurities like sulfur and iron-containing matters [3]. The second stage, conducted with a crucible pot supplied with a stirrer, outputs a strong solution containing 500-520 g NaOH/L. The SHA precipitation progresses in a water-cooled crystallizer. The filtered solids are packed in polyethylene bags. The SHA produced is a white

92

loose substance with a composition of 31-34% NaOH, 2% Na_2CO_3, 34-40% Al_2O_3, 0.2% SiO_2, and 0.2% Fe_2O_3. The moisture content of the product (14-20 wt. %) impedes dust formation while transferring operations. An alternative method of wastewater utilization is its drying in a fluidized bed granulator (FBG). The initial spent liquor is evaporated to 390-430 g NaOH/L, and is then corrected with an alumina hydrate addition to α_{caust} =1.5-1.6, clarified, and collected in a service tank. From here, it is pump-sprayed through a pneumatic nozzle in FBG (Figure 3.2.6).

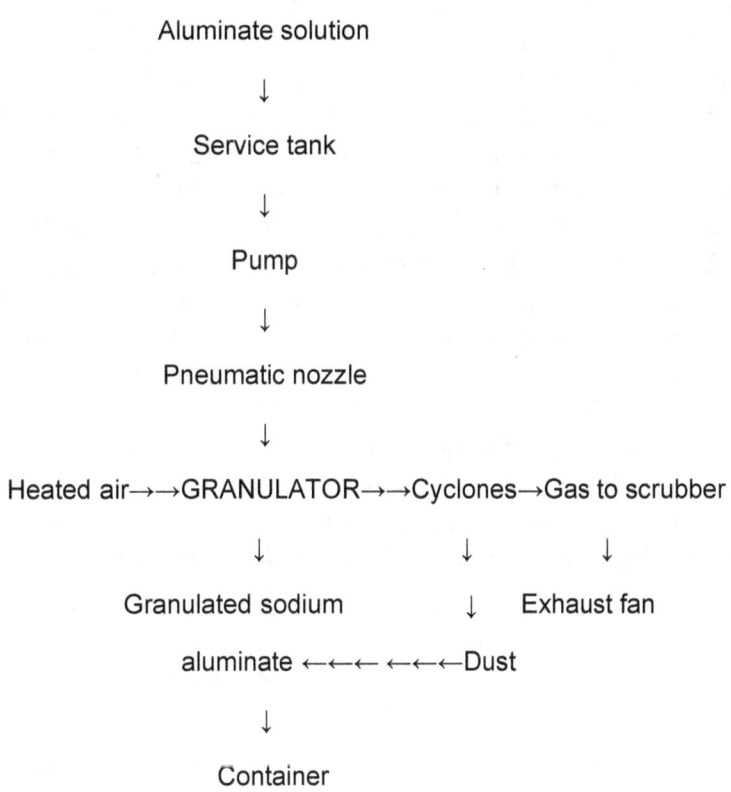

Figure 3.2.6. A schematic sketch of a processing plant for drying and granulating SA produced from pre-evaporated wastewater

At the same time, air that is fed under the furnace bottom plays the role of a fluidizing agent and heat carrier. It is forced by a fan and heated by fuel gases sequentially in convective and radiation recuperators. Dusty air becomes clean in coupled cyclones and scrubber. The FBG is operated under a continuous solution feed and synchronous granules discharge, meaning that the fluidized bed is practically isothermal. Hence, the temperatures of both granules and over-bed air are the same (390-400°C). The temperature of the heated air is 750-800°C; temperature of the bunkered granules is 300-320°C. The velocity of the air in the bed is 2.1 m/s and is 0.15 m/s over the bed, while the velocity of the injected solution is 25-27 m/s. The bed resistance is 0.04-0.45 MPa. Moisture removal is

1,200-1,250 kg/m^2.h, and the percentage of dust carried away is 5-7%. The average grain-size granules distribution is (wt.%) -20 + 10 mm = 0.4, -10 + 5mm = 9.1, -5 + 3mm = 32, -3 + 1mm = 47.5, -1 + 0.5mm = 6, and - 5mm = 4.3. Thus, about 85 % of product-forming particles are sized between 5 and 0.5 mm. The SA commodity contains, on the average, 42.9% Na$_2$O as a part of NaOH, NaAlO$_2$, and Na$_2$Al(OH)$_5$ as well as 43.5% Al$_2$O$_3$, 4.7% Na$_2$CO$_3$, and 8.9% LOI. New aluminate- containing wastes have been use in manufacturing SA and SHA at the above described facilities to meet consumers' requests. There is about 28m^3 of wastewater per ton of final product run off in the manufacturing of phenylethanol (PE) [43]. The waste contains 39-45 g/L NaOH, 30-35 g/L Al$_2$O$_3$, 90 g/L NaCl, and 2-2.5g/L HOCH$_2$ x CH$_2$OH (glycol). It was utilized in an fluidized bed granulator (FBG)[44]. As wastewater is evaporated before drying and SA granulation, the NaCl concentration rises initially, but then decreases due to sodium chloride salting out. When the NaOH concentration attains its end point (415-420g/L), the NaCl content equals 79-80g/L, and, in doing so, the NaAlO$_2$:NaCl ratio is changed from 0.666:1 in the initial wastewater to 8.1:1 in the evaporated solution. The glycolic part of solution feeding into the FBG is completely burnt in the course of drying. Granules of obtained SA have a white color and contain 37% Na$_2$O, 35% Al$_2$O$_3$, and 16% NaCl. The etch-bath wastewater is another source for sodium aluminate [44,45]. It contains 44-47g/L NaOH and 35 g/L Al$_2$O$_3$, as well potassium, thiosulfate, and some fluorine. The initial wastewater is pre-evaporated to an NaOH concentration of 250-260 g/L. For SHA, obtaining this solution is subjected to an additional in-depth evaporation, followed by cooling. The precipitated product is filtered and packed. The process data are given in Table 3.2.3.

Table 3.2.3. The change in composition of solutions in the course of processing PE-manufacture wastewater into SHA.

Solution	NaOH	Al$_2$O$_3$	Concentration (g/l) SO$_4$$^{2-}$	S$_2$O$_3$$^{2-}$	KOH
Initial	256.8	186.1	-	21.3	5.9
Evaporated	440.5	310.0	29.3	9.8	-
Depleted	419.6	272.9	20.9	7.9	-

The 440g/L NaOH concentration of evaporated solution is significantly lower than the 500-555 g/L NaOH concentration required in the relatively pure solutions produced from the alumina-refinery wastewater. The difference is due to salting-out impact of potassium and sulfur-bearing dissolved compounds that exceed concentrations of 35 g/L in the evaporated solution. The dry part of the product obtained contains 31-32% NaOH, 27% Al$_2$O$_3$, 6-6.5% Na$_2$SO$_3$ + Na$_2$SO$_4$, 1.6-1.7% K$_2$O, and 0.1% F. The filtered cake moisture is 20-25%. About 55-60% of dissolved sodium aluminate is precipitated and filtered with a specific productivity of 120-135 kg/m^2.h. The same pre-evaporated solution to be dried in an FBG is finished to a concentration of 390 g/L NaOH and sprayed in a granulator. Granules are between 1 mm and 10 mm and have a bulk weight of 1,150-1,175 kg/m^3 and a water solubility of 98-100%. The product chemical composition is

45.1% Na_2O, 41.9% Al_2O_3, 1.8% CO_2, 1.9% SO_3, and 0.7% K_2O. Both SA and SHA produced by wastewater utilization were profitably employed as constituents of drilling mud solutions, additives to early-strength concrete, and coagulants in paper manufacture [3,45,46]. Apart from SA-containing wastewater, the mass of alumina-bearing solid chemical refuses are not currently processed, but could be used.

Sodium Aluminate Manufacturing from Solid Wastes. The most abundant alumina-bearing solid wastes formed directly in the aluminum industry are spent potliner [47] and salt cake [48]. There are feasible avenues for obtaining SA from both wastes. Spent potliner (SPL) is considered a hazardous waste due to the presence of cyanides and fluorides. The amount of SPL generated worldwide amount is estimated to be close to one million tones [49]. SPL composition may be changed over a wide range. As an example, the average elemental composition of cathode waste is [50] 33.1% C, 15.7% F, 15.1% Al, 14.2% Na, 2.7% Si, 1.8% Ca, 0.3% CN, 0.1% S, and 17% O_2 + trace materials. Any process using caustic or soda may result in sodium aluminate manufacture. In addition to the techniques mentioned [49,51], a method was developed to form a combined sodium aluminate and sodium fluoride solution (Figure 3.2.7) [52]. Raw material is digested by an alkali-aluminate solution taken from the Bayer process cycle; dump sludge containing iron, silicon, and other impurities; and aluminate-fluorid loquor. The latter is pre-evaporated to 320-400 g NaOH/L; the obtained solution may be transferred into a solid state either by drying and granulation or by finishing evaporation and crystallization. The first product is sodium-fluoric aluminate (SFA); the second resulting material is sodium-fluoric hydroaluminate (SFHA). The impact of sodium hydroaluminate precipitation of fluoride in aluminate solutions on was studied [53]. It was found that fluoride acts as a catalyst and crystal enlarger. This results in a velocity increase in seeds formation, followed by structure ordering. The initial alkali-aluminate solution containing 580 g NaOH/L 106 g Al_2O_3/L and 2 g SiO_2/L was evaporated to an NaOH concentration of 685-700g/L. The pre-precipitated SHA residue was used as a seed and put in evaporated solution before cooling in amounts of 80% of Al_2O_3 contained in the liquor. Fluorine concentration was varied between 0.1-0.4 g/L.

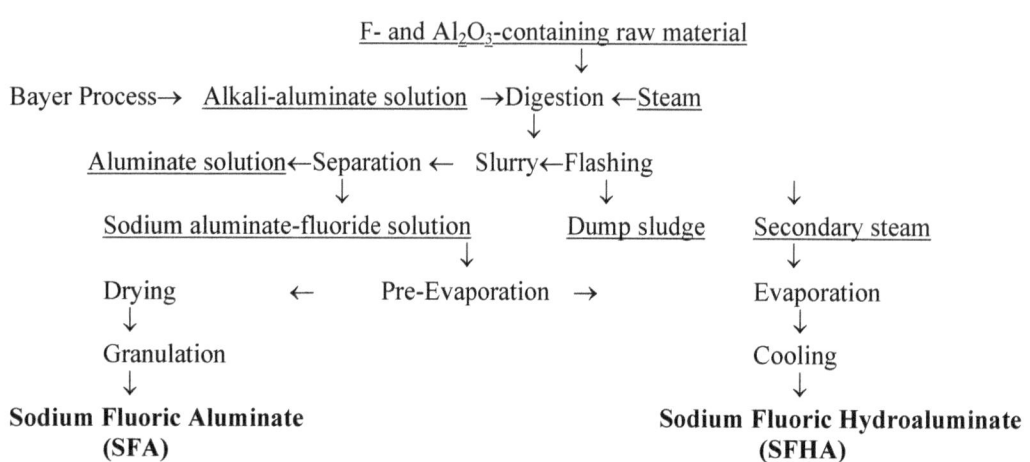

Figure 3.2.7. A flow chart for hydroalkaline processing alumina and fluorine-bearing raw material into sodium fluoric aluminate and hydroaluminate

The evaporated solution was cooled to 45°C and stirred for eight hours. The concentration of Al_2O_3 was controlled during the test time (Figure 3.2.8).

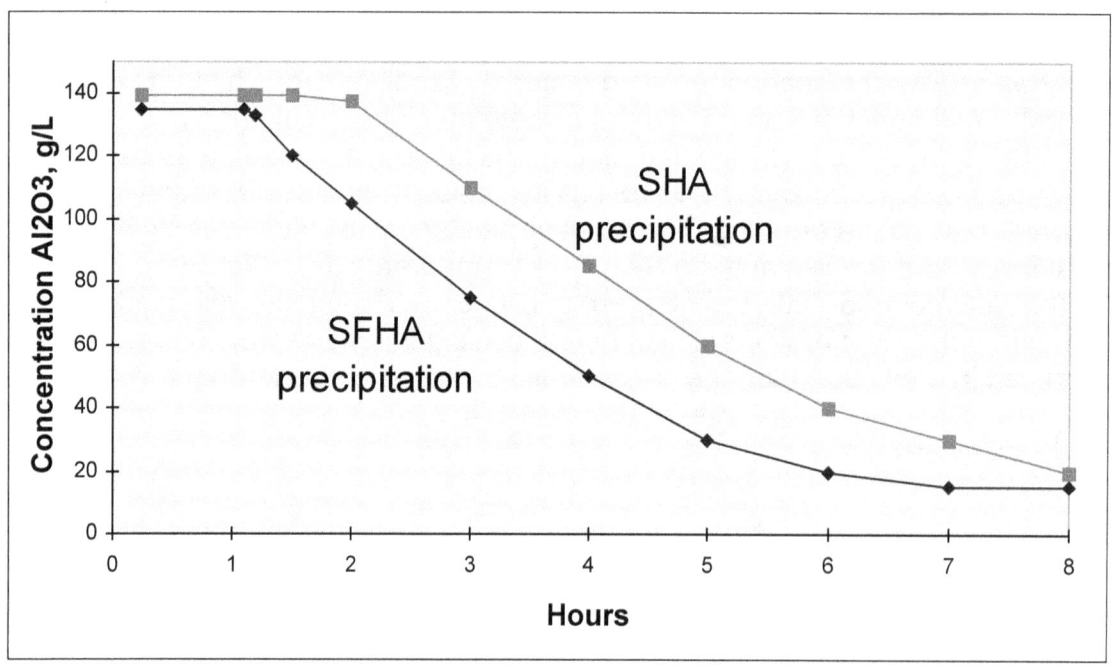

Figure 3.2.8. The kinetics of SHA and SFHA precipitation from alkali-aluminate solution at 45°C

The induction time for SFHA crystallization is less than that for fluorine-free SHA.

96

In the first case, precipitation was completed two hours before it occurred in the fluorine absence. Due to the F-addition, $\alpha_{caust.}$ of the precipitated product was reduced from 1.7 to 1.26, with the Al_2O_3 yield increased to 85% as a result of the formation of a coarsely crystalline residue (particle size is equal to 100μ and more).The use of SFHA in combination with aluminum sulfate for drinking-water purification results in a less intensive fluoride decrease in cleaned water as compared to the use of aluminum sulfate alone [54]. The maximum drop in F concentration is 17% for the combined reagent and 30% for $Al_2(SO_4)_3$. The combined use of aluminum sulfate and SFHA allows for simultaneous purification and fluoridation of the water. The optimal ratios of $Al_2(SO_4)_3$ to SFHA are (0.25-0.1):1 and (0.7-0.4):1 at an initial water turbidity of less than 70 mg/l and greater than 70 mg/l, respectively. The fluoridation efficiency depends on the order of coagulants addition to the water. The best result is achieved if SFHA is added first, and aluminum sulfate is added 40-60 minutes later. Alumina-bearing salt cake is a byproduct of aluminum dross utilization. Its reclaiming and reuse is an important problem of aluminum-reduction technology. World dross and salt cake generation by aluminum recycles amounts to about 2.5-3.5 million tones per year [55]. Dross formation occurs during used beverage cans melting under a protective molten salt flux cover. The flux prevents metal oxidation, provides coalescence of the liquid aluminum droplets, and separates recovered metal from impurities. The dross is a mixture of Al_2O_3, salt, metallic aluminum, and nonaluminic residue. An example of the salt-cake recycling process flow chart is shown in Figure 3.2.9a [56].

The crushed salt cake feeds to a tank where it is water leached. Salt is dissolved to a brine concentration of 22 wt. % salt. The solid remainder, which is mostly composed of aluminum oxide, is removed as a wet residue and landfilled. The separated brine liquor is evaporated, and both sodium and potassium chlorides are crystallized. This process, as well as other versions, do not involve the conversion of wet residue into alumina or alternative products. On the basis of salt cake composition, including up to 50-85% Al_2O_3 [56], more than 200,000 tonnes of alumina are lost annually in such a manner. At the same time, all or part of this amount can be processed into specific SA and/or SHA. To do this would require the interaction $Na_2O + Al_2O_3 = Na_2Al_2O_4$ to occur at some stage of the salt-cake recycling procedure. In this case, the final part of the flow chart in Figure 3.2.9a would be modified (Figure 3.2.9b). One possible process to obtain SA and SHA was tested on a bench scale. The obtained dry sodium-saline aluminate (SSA) contained as much as 35.9% SA and 60.1% of both sodium and potassium chlorides. NaCl solubility in aluminate solutions was also studied [57]. The dependence of NaCl concentration on the sodium oxide content for a solution of

α_{caust} =2 is diagrammed in Figure 3.2.10.

It is evident that solution designated for processing into SSA should be evaporated to about 330 g NaOH/L and 70 g NaCl/L. The final concentrations of NaOH and NaCl in the higher concentrated liquor would be 520 g/L and 10 g/L, respectively. The recycled sodium chloride would precipitate before sodium hydroaluminate is crystallized. The crystal hydrate recovered from aluminate Cl-containing solutions is identified as a sodium chlorohydroxoaluminate (SCHA) [58]. On the basis of X-ray diffraction, infrared spectral and thermal analysis data, the chemical formula for SCHA is $Na_{1.13}Al[O_{2.22}(OH)_{1.75}Cl_{0.24}]_{0.64}$. Its structure relates to the prismatic space group P_{mmm} or F222 or $F_{mm}2$, **a** 10.45, **b** 11.57, and **c** 12.43nm. A similar product has been manufactured in the course of wastewater utilization. The chlorine-containing sodium aluminate was adequately used in paper-making, where it must be in large demand.

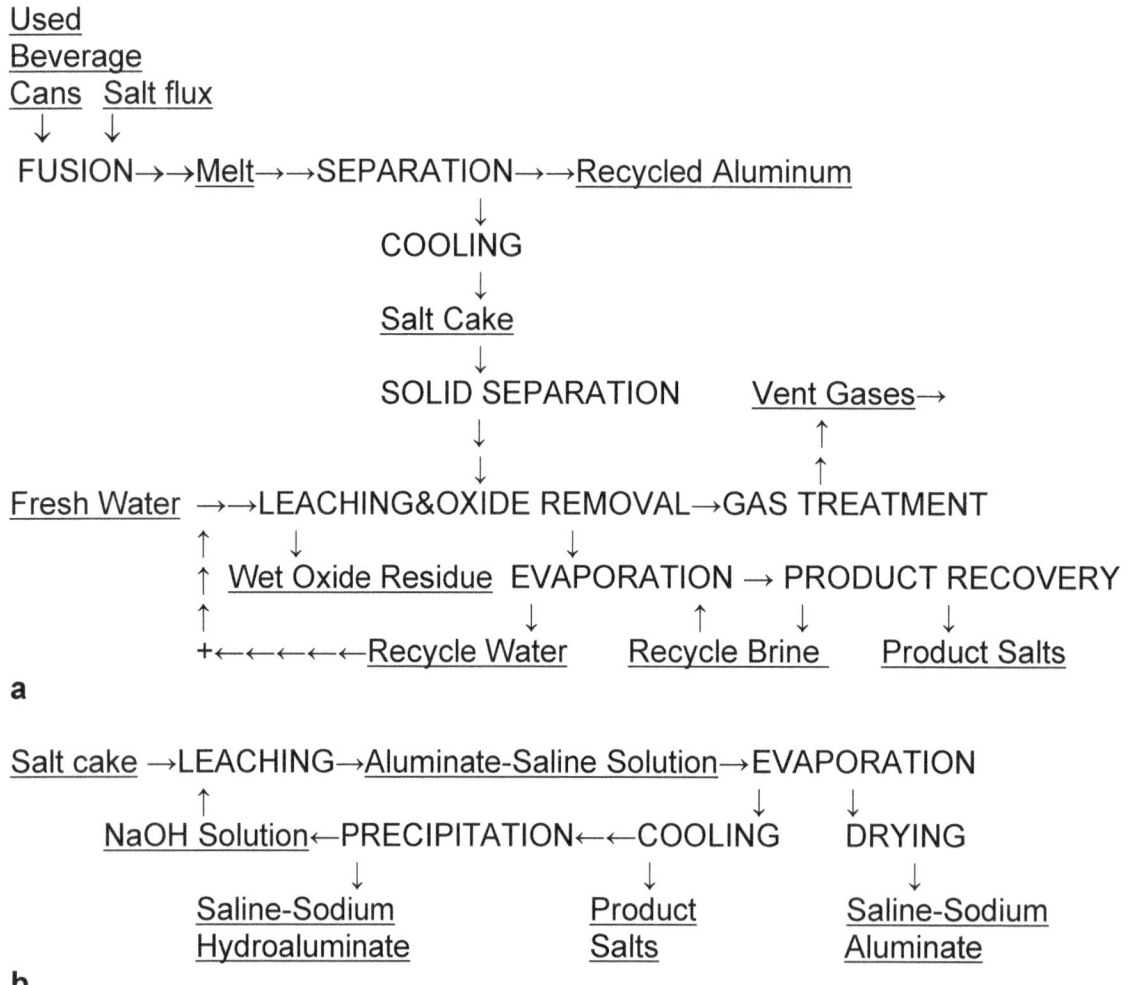

Figure 3.2.9. A flow chart of (**a**) the basic salt-cake recycling process and (**b**) the process modified for SSA ans SSHA

98

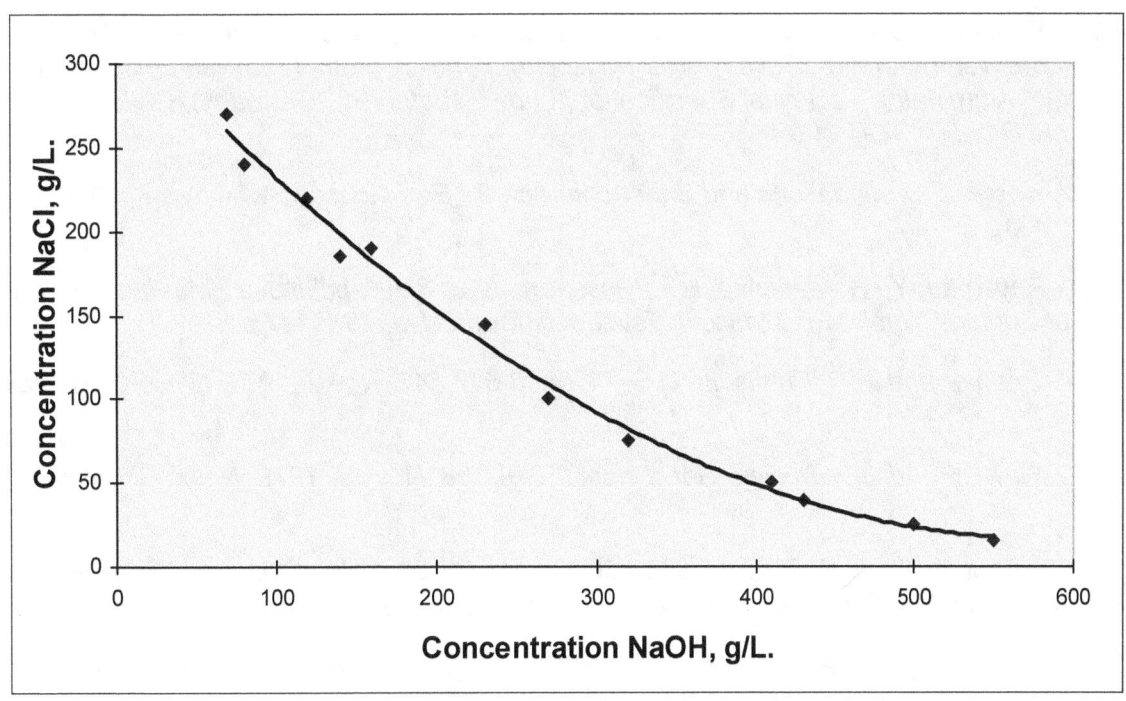

Figure 3.2.10. The relationship between NaCl solubility and NaOH concentration for aluminate solution (molar ratio of $Na_2O:Al_2O_3=2:1$).

Conclusion. Many hundred thousands tonnes of SA and SHA may be annually manufactured by utilization of spent potliner and salt cake. The well-founded theoretically, developed and industrially tested processes would provide to turn out both products in dry, wet (as crystallohydrate) or liquid form. The impurities presence, such as F^- and Cl^-, must not impair the materials suitability. Apparently, the worldwide requirements in these chemicals do not exceed one or two hundred thousands tonnes per year. However, there would be no difficulties in processing the excessive sodium aluminate-containing products into alumina, caustic, soda ash and other valuable products. Thus, huge mass of alumina-bearing wastes will also find an expedient application.

References:

1.Kirk-Othmer Encyclopedia of Chemical Technology, 4th Ed., New York, John Wiley & Sons, Inc., 1992.

2.Production of Aluminum and Alumina, Ed. A.R. Burkin, New York, John Wiley & Sons, Inc. 1987, 241 p.

3.V.L. Rayzman, Yu.K. Vlasenko, L.S. Nisse, E.V. Vilson et al., "Sodium Aluminate and Hydroaluminate: Production and Use", Rostov-na-Donu (Russia), Rostov University, 1990, 120 p.

4.V.L. Rayzman, S.N. Linevich, E.V. Vilson, et al., "Application of Aluminate Solutions to Clean the Natural Water", Tsvetn. Met. (Moscow), 1993, 3, p. 36-37. 5.Natriumaluminat (Sodium Aluminate), Technical Specification, Ludwigshafen/Rh (Germany), Glulini-Chemie GmbH, 1979, 14 p.

6.E.V. Kosterin, "Foundations and Substructures", 3rd Ed., Moscow, Vyshaya Shkola, 1990, 431p.

7.V.L. Rayzman, Yu.K. Vlasenko, L.S. Nisse, and N.V. Sinel'shchikova, "Manufacture and Use of Sodium Aluminate", Moscow, Tsvetmetinformatsiya, 1987, 47 p

8.R.A. Boehler and M.R. Purvis, Jr. U.S. Patent 3,617,542, Nov 2, 1971, Nalco Chemical Co.

9.R.D. Sawyer and J.D. Tinsley, Ger. Offen. 2,016,758, Nov 19, 1979, Nalco Chemical Co.

10.D.G. Braithwaite, U.S. Patent 2,996,460, Aug 15, 1960, Nalco Chemical Co. 11.R.M. Milton, U.S. Patent 2,882,243, Apr 14, 1959, Union CarbideCorp.

12.A.K. Zapol'skiy and A.A. Baran, "Coagulants and Flocculants in the Water Purification Processes", Leningrad, Khimiya, 1987, 208 p.

13.R. Fricke and P. Jucaitis, "Untersuchungen uber die Gleichgewichte in den System Al_2O_3-Na_2O-H_2O und Al_2O_3-K_2O." Z. Anorg. Allgem. Chem. 1930, B. 191, S. 129-149.

14.P. Jucaitis. "Uber die Zusammenserzung und Konstitution der Alkalialuminate", Z. Anorg. Allgem. Chem. 1934, B. 220, S. 257-26

15.I.W. Sprauer and D.W. Pearce, "Equilibria in the Systems Na_2O-SiO_2-H_2O and Na_2O-Al_2O_3-H_2O at 25ºC", J. Phys. Chem. 1940, 44, p. 909-916.

16.S.I. Kuznetsov and V.A. Derevyankin, "Physical Chemistry of the Bayer Process for Alumina Refinery", Moscow, Metallurgy, 1964, 352 p.

17.G.A. Panasko and P.V. Yashunin, "On Calculations of Na_2O-Al_2O_3-H_2O System", Journal Appl. Chem. (USSR), 1964 ,2, p. 285-289.

18.W.L. Masterton, E.J. Slowinski, and C.L. Stanitski, "Chemical Principles", New York, Saunders College Publishing, 1983, 791p.

19.V.L. Rayzman and Yu.K. Vlasenko, "Ionic Strength and Stability of Aluminate Solutions", Tsvetn. Met. (USSR) 1988, 11, p. 57-59.

20.V.L. Rayzman, "Determination of the Thermodynamic Quantities of Aluminate Ions", Tsvetn. Met. 1985, 5, p. 60-62.

21.V.L. Rayzman, "Calculation of Heats of Formation of Molecules and Ions of Complex Composition in Aqueous Solutions", Zh. Obshch. Khim., (USSR) 1985, 55(3), p. 503-508.

22.V.L. Rayzman, L.P. Ni, Yu.K. Vlasenko, and V.I. Pevzner, "A Study of Boundaries Between Stability Fields of Different Aluminate Ions in the System Sodium Oxide-Alumina-Water", Kompleksn. Ispol'z. Miner. Syr'ya (USSR), 1986, 7, p. 61-65.

23. V.L. Rayzman, Yu.K. Vlasenko, V.S. Sazhin, and I.Z. Pevzner, "Thermodynamic Analysis of Interactions in the High-Alkali Part of the Sodium Oxide-Alumina-Water System", Tsvetn. Met. (USSR), 1982, 8, p. 36-38.

24. V.L. Rayzman, Yu.K. Vlasenko, S.P. Gabuda, and A.M. Panich, "Investigating the Structure of Sodium Aluminate by the NMR Technique", Izv. Akad. Nauk USSR, Neorg. Mater. (USSR), 1985, 21(12), p. 2065-2068.

25. V.L. Rayzman and Yu.K. Vlasenko, "Study of Decomposition of Highly Concentrated Aluminate Solutions", Proceedings "New Solutions for Apparatus and Processes Development in Production of Alumina, Aluminum and Intermediate Materials, Leningrad, VAMI, 1986, p. 12-15.

26. A.I. Lainer, N.I. Eremin, Yu.A. Lainer, and I.Z. Pevzner, "Alumina Production", Moscow, Metallurgy, 1978, 344 p.

27. V.L. Rayzman, L.S. Nisse, and Yu.K. Vlasenko, "Pilot Plant Test of Technology for Purification of Nikolaev Alumina Plant Aluminate Solutions from Impurities through Sodium Aluminates Production", Research Reportf, Leningrad, VAMI Pilot Plant, 1988, 52 p.

28. E. Asselborn, "Nepheline" (Mine'r. et fossils), Montreal, 1983, 9 (102), p. 41-42.

29. G.A. Panasko and M.N. Smirnov, "Recovery of Sodium Caustic from Combined Aluminate Solutions", Leningrad, Proceedings of VAMI, 1970, 70, p. 126-135.

30. V.L. Rayzman, L.S. Nisse, Yu.K. Vlasenko, and E.I. Andreev, "Crystallization of Alkaline Hydroaluminate from Low Modules Sodium-Potassium Solutions", Issled. Tekhnol. Protsessov Proizvodstva Glinozyoma iz Razl. Vidov Glinozyomsoderzh. Syr'a, Leningrad, Proceedings of VAMI, 1989, p. 59-61.

31. V.L. Rayzman, Yu.K. Vlasenko, L.S. Nisse et al., "Method for Obtaining Alkali Metal Hydroaluminate", PCT Int. Appl. WOC 8701,107, Feb 26, 1987.

32. V.L. Rayzman, Yu.K. Vlasenko, and V.M. Sizyakov, "Theoretical Premises to Preparation of Sodium Hydroaluminate from Low-Modules Solutions", Izv. Vyssh. Uchebn. Zaved., Tsvetn. Metall., 1986, 5, p. 46-50.

33. V.L. Rayzman, Yu.K. Vlasenko, and L.S. Nisse, "Study of Crystallization of Sodium Aluminate from Low-Modules Aluminate Solutions", (USSR) Kompleksn. Pererab. Glinozyomsoderzh. Syr'ya, Leningrad, Proceedings of VAMI, 1987, p. 43-47

34. V.S. Sazhin, "New Hydrochemical Procedures for Comprehensive Processing of Aluminosilicates and High-Silicon Bauxites," Moscow, Metallurgy, 1988, 215 p.

35. H.J. Hittner, "Hydrothermal Alkaline Process for Extraction of Alumina from Anorthosite", Trav. Com. Int. Etude Bauxite, Alumine et Alum., 1981, 16, p. 13-21.

36. V.L. Rayzman and I.K. Filipovich, "Study of the Phase Composition of the Autoclave Nepheline Sludge", Issled. Tekhnol. Protsessov Pr-va Glinozyoma iz. Razl. Vidov Glinozyomsoderzh. Syr'ya, Leningrad, Proceedings of VAMI, 1989, p. 62-64.

37. K. Hudson and T.G. Swansiger, "Recovery of Sodium Aluminate from High-Silica Aluminous Materials" US 3,998,927, Dec. 21, 1976, 6 p.

38. H. Ivekovic and Z. Balenovic, "Sodium Aluminate from Aluminate Layers", Proceedings of "ICSOBA" (Budapest: 1969), pp. 103-107.

39. V.L. Rayzman, V.V. Volkov, J.B. Rosen et al., "Method for Processing Alumina-Bearing Raw Material", S.U patent 1,508,530 (appl. Aug 17, 1987).

40. N.N. Tikhonov, V.V. Volkov, V.L. Rayzman et al., "To Develop and Master at Semiindustrial Scale the Method for High-Temperature Hydrochemical Processing of Low-Grade Aluminum-Containing Raw Material. Chapter T2. To Develop and Test the Process under Conditions of Leningrad "VAMI" Pilot Plant", Research Report, Leningrad, VAMI, 1990, 148 p.

41. V.L. Rayzman, G.V. Smirnova, I.K. Filipovich et al., "Processing of Lithium-Containing Alumino-Alkaline Solution", SU patent (appl. March 28, 1988, approved).

42. J.B. Rosen, V.A. Volkov, V.L. Rayzman et al., "Desilication of Wastewater of Leningrad VAMI Plant", Tsvetnaya Metallurgiya, 1980, 10, p. 32- 33.

43. M.L. Vorob'yov, ed., Methods for Processing and Feasible Avenues for Industrial Use of Aluminum-Containing Wastes", Proceedings of Scientific and Technical Symposium Scientific & Technical Society, Gorki (Nizhniy Novgorod), 1983, 35 p.

44. A. Fendel and J.Lehmkuhl, "Wastewater Treatment and Recovery from Etch Bath and Other Residues in the Surface Treatment of Aluminum", Aluminum (Int. J. for Ind., Res. & Appl.), 1997, 73 (6), p. 408-413.

45. J.B. Rosen, V.L. Rayzman, L.S. Nisse et al., "Processing Aluminum-Containing Wastes into Sodium Aluminate", Tsvetnaya Metallurgiya, 1984, 4, p. 41-44.

46. A.M. Privalov, L.S. Nisse, V.L. Rayzman, and J.B. Rosen, "Preparation of Salt Solution of Aluminum Production to Utilization", Tsvetnaya Metallurgiya, 1987, 4, p. 46-48.

47. A.J. Plumpton, J.-F. Wilhelmy, D. Blackburn, and J.-L. Caouette, "Mineralogical and Physical Considerations Related to the Separation and Recovery of Constituents from Aluminum Smelter Byproducts and Wastes, " Light Metals 1996, ed. Wayne Hale, Warrendale, PA, TMS, 1996, p. 1271-1273.

48. G.J. Kulik and J.C. Daley, "Aluminum Dross Processing in the 90's," Proceedings of the 2nd International Symposium on Recycling of Metals and Engineering Materials, eds. J.H.L. van Linden, D.L. Stewart, Jr., and Y. Sahai, Warrendale, PA, TMS, 1990, p. 427-437.

49. R.P. Pawlek, "Recent Developments in the Treatment of Spent Potlining," JOM, November 1993, p. 48-52.

50. L.G. Boxall, "Cathode Waste Disposal Options," 2nd Australasian Aluminum Smelter Technology Course, Sydney, Australia, 1987, p. 21-1 – 21-15.

51. R.J. Adrien, J. Besida, T.K. Pong et al., "A Process for Treatment and Recovery of Spent Potliner," Light Metals 1996, ed. Wayne Hale, Warrendale, PA, TMS, 1996, pp.

1261-1263.

52. N.S. Volkova, M.N. Smirnov, V.M. Sizyakov, I.Z. Pevzner, and V.L. Rayzman, "Process for Cryolite Production," SU patent 1,034,341, Apr 08, 1983.

53. L.P. Ni, E.V. Starshikova, R.A. Abdulvaliev, V.L. Rayzman, and V.V. Medvedev, "Development of the Procedure of Sodium Hydroaluminate Crystallization from High-Modules Solutions", Proceedings of Symposium of Scientific and Technical Society "Up-to-date Concepts of Constitution of Alkali-Aluminate Solutions and Processes for Precipitation of Aluminum and Sodium – Containing Compounds from these Solutions," Sverdlovsk, Scientific and Technical Society, Sep 30-Oct. 03, 1986, p. 27-28.

54. V.L. Rayzman, S.N. Linevich, E.V. Vil'son et al., "Use of Fluorine-Containing Sodium Hydroaluminate in Reagent Treatment of Drinking Water," Tsvetn. Met., 1993, 2, p. 30-31.

55. R.P. Pawlek, "Recycling", Aluminum, 1997, 73 (6), p. 419.

56. D. Graziano, J.N. Hryn, and E.J. Daniels, "The Economics of Salt Cake Recycling," Light Metals, 1996, ed. Wayne Hale, Warrendale, PA, TMS, 1996, p. 1255-1260.

57. J.B. Rosen and V.L. Rayzman, "Study of Solubility of Sodium Chloride in Aluminate Solutions," Tsvetn. Metal., 1989, 1, p. 65-66.

58. O.B. Khalyapina, V.L. Rayzman, L.P. Ni et al., "Chlorine-Containing Sodium Hydroaluminate," Izv. Akad. Nauk USS, Neorg. Mater., 1991, 27 (9), p. 1891-1894.

3.3. Stability of Alkali-Aluminate-Silicate Solutions in the Na_2O-Al_2O_3- SiO_2-H_2O System.

Alkali-aluminate-silicate solutions are a section of the system Na_2O (K_2O) –Al_2O_3-SiO_2-H_2O. Interactions in this system have provided a technological basis for combined processing high-silica alumina-bearing raw material, including ore conditioning and also precipitation of silica from alkaline aluminosilicate solutions in the composition of sodium or potassium hydroaluminosilicates (SHAS or PHAS). Efficiency of the solution desilication process can be ensured by establishing conditions such that the equilibria between the sodium or potassium cations and the hydroxyl aluminate, and silicate anions will be shifted toward the formation of SHAS and PHAS. The same time, for maximum extraction of silicon into the solution in the process of conditioning high-silica raw material, it is necessary to stabilize the aluminate-silicate solution that is formed, so that it can be separated from the concentrate without premature passage of the silicate ion into the solid residue, thus obtaining a silicon-lean solid commodity. In Reference [1], in describing the kinetics of desilication of aluminate solutions, the existence of "concentration plateaus" was noted, these plateaus being characterized by a high (at least 5 g/L) and stable content of SiO_2. The present work has been aimed at a theoretical justification for the stability of the silicate ion within the framework of the Na_2O-Al_2O_3- SiO_2- H_2O system; this is of great importance in terms of increasing the efficiency of aqueous-caustic conditioning low-quality alumina-bearing raw material. It appears to us that thermodynamic analysis of the equilibrium of SHAS with an alkaline aluminate-silicate solution is the most objective method for solving the problem at hand, eliminating the performance of extended, repetitive experiments. As the initial thermodynamic premises the data obtained by Kosova and Dem'yanets [2] may be taken. The authors investigated the interaction of one of the widely encountered vareties of SHAS-hydrocancrinite, $Na_8[AlSiO_4]_6(OH)_2.2H_2O$ with an NaOH solution at temperatures of 200°C, 250°C, and 300°C. The values of the constants of SHAS interaction with the sodium hydroxide solution K_T^o within this temperature interval and with solution ionic strength 0.8-5.3, as obtained in Reference [2], indicate that this process goes from right to left, i.e., in the direction of SHAS formation (log K_T^o < 0). The authors of Reference [2] found the value of log K_T^o in the following expression:

$$y = - \log K_T^o + pl \qquad (1)$$

where y is a complex function of the ionic strength I of the solution, as determined by the Debye-Huckel equation;

p is adimensionless proportionality coefficient with a value of 0.753 at 200°C, 0.703 at 250°C, and 0.929 at 300°C.

Plots of data on the coordinates of Equation (1) are shown in Figure 3.3.1.

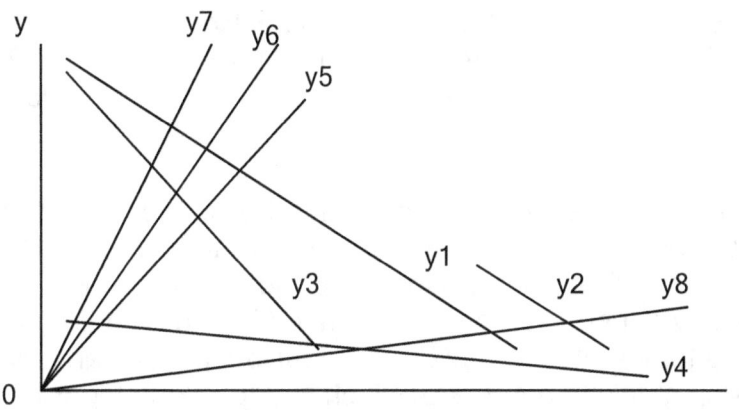

Figure 3.3.1. Plots of y=(-log $K°_T$) + pl for indicated temperatures (°C):
200 (y_1), 250 (y_2), 300 (y_3), 65 (y_4).
Plots of y=pl with log $K°_T$=0 for following temperatures(°C):
200 (y_5), 250 (y_6), 300 (y_7), 65 (y_8).

The intercepts of lines 1-3 on the vertical axis are numerically equal to – log $K_T°$. With – log $K_T°$ = 0, y = pl. Consequently, the points of intersection of lines 1-3 with rays 6-8 determine the values of the critical ionic strength of solution (I_{cr}) at which the reaction of desilication of the solution changes its direction toward dissolution of the hydrocancrinite. These values of I_{cr} are as follows: 19.5 at 200°C; 20.5 at 250°C; and 15.5 at 300°C. The ionic strength of a solution is understood to be the half-sum of products of molarities m of all ions by the square of their valence. In our case, the expression for I has the form

$$I = 0.5[m_{OH^-} + m_{Na^+} + m_{AlO2^-} + 4m_{(H2SiO4)2^-} + m_{H3SiO4^-} + 16m_{(SiO4)4^-}] \qquad (2)$$

By transformation of Equation (2) in terms of the solution parameters – density, sodium oxide concentration [Na_2O] (g/L) – and the concentration ratios

α = Na_2O:Al_2O_3 (mole) and μ = Al_2O_3:SiO_2 (mass), the equation is obtained

$$I = \boldsymbol{a} (1 + \boldsymbol{b}/\alpha\mu) : /0.0103 - 1/[Na_2O] - \boldsymbol{c}/\alpha - \boldsymbol{d}/\alpha\mu/ \qquad (3)$$

By substituting into Equation (3) the values of I_{cr} corresponding to the temperatures 200°C, 250°C, and 300°C, and then comparing various combinations of the parameters [Na$_2$O], α, and μ, using a special "VAMI"'s computer software, the values of the coefficients a, b, c, and d were obtained (Table 3.3.1).

Table 3.3.1. The values of the coefficients of Equation (3) at some temperatures

Temperature (°C)	a	b	c	d
200	$3.62 \cdot 10^{-2}$	0.189	$-5.25 \cdot 10^{-2}$	-0.311
250	0.154	25.792	$1.141 \cdot 10^{-3}$	-0.221
300	0.123	$1.141 \cdot 10^{-13}$	0	$9.055 \cdot 10^{-16}$

Under production conditions, procedures of separating the solution from the solid residue are carried out at temperatures generally no higher than 100°C. The possibility of maintaining stability of the solution under "low-temperature" conditions was checked by a thermodynamic calculation based on data obtained by Pevzner et al [3] for a temperature of 65°C. The equilibrium constant-solution ionic strength relationship at this temperature, which is depicted in Figure 3.3.1 (I_{cr} /65°C/ = 23.7), is characterized by the following magnitudes of the coefficients of Equation (3): a = 0.183, b = 437.874, c = $-2.673 \cdot 10^{-3}$, d = -3.432. Using the thermodynamic data that have been obtained above, the lower concentration limits of stability for alkaline aluminosilicate solution in the Na$_2$O-Al$_2$O$_3$- SiO$_2$- H$_2$O system were calculated; these are shown in Figure 3.3.2. Thus, condition (3) reflects the phenomenon of coexistence of chemically active ions in solutions of the Na$_2$O-Al$_2$O$_3$- SiO$_2$- H$_2$O system without the formation of any solid-phase compounds, this phenomenon consisiting of stabilization of ions when the ionic strength of the solution becomes sufficient to ensure that the chemical reaction will proceed toward dissolution of the solid phase.

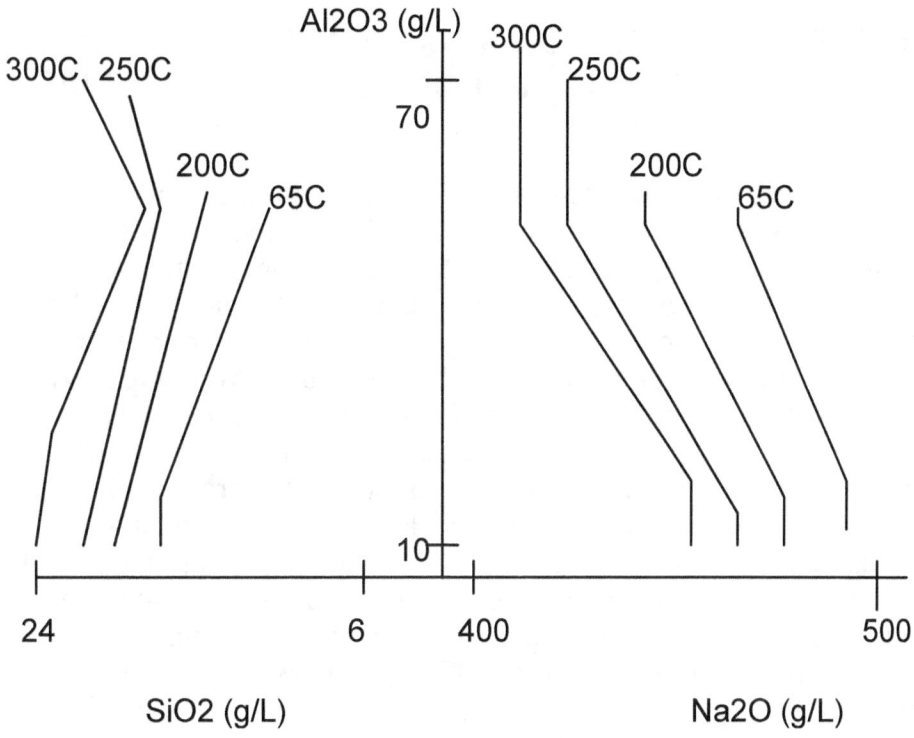

Figure 3.3.2. Calculated lower concentration limits of stability for solutions in Na_2O-Al_2O_3-SiO_2-H_2O system at temperatures 65°C, 200°C, 250°C, and 300°C

References:

1.L.P. Ni and V.L. Rayzman, "Combined Methods for Processing Low-Quality Aluminum-Containing Raw Materials," Alma-Ata, Nauka, 1988, 288 p.

2.T.B. Kosova and L.N. Dem'yanets, in: "Growth of Crystals in High-Temperature Aqueous Solutions," Moscow, Nauka, 1977, p. 43-65.

3.I.Z. Pevzner, N.I. Eryomin, J.B. Rozen, et al., Zh. Prikl. Khim., 47 (3), 1974, p. 2758-2760.

Chapter 4. INTEGRATING COAL COMBUSTION AND PROCESSES FOR RECOVERING ALUMINA FROM INDUSTRIAL WASTE

4.1. About Co-processing Coal Ash and Red Mud into Alumina and Byproducts. The fuel cost at alumina refinery can be significantly reduced by shifting a steam generation from gas or oil to coal use. The sorbent that can be applied for the cleaning of fuel gases is a regular red mud copiously dumped at alumina refineries. It is possible to combine coal combustion and red mud low-temperature sintering to recover remaining alumina and caustic from a red mud. The integrated technology, based on a combination of known processes will provide advantages over other methods that are currently being used. Depending on location, coal is a cost-effective energy source for steam generation that reduces the cost of energy for producing alumina. However, most alumina refineries employ oil or natural gas for this purpose. The present market oil price is 30% to 50% of what alumina can be sold for, while coal prices are between 10% and 18% [1]. Thus, the impact of converting oil or gas to coal is significantly greater than other attempts to improve energy efficiency. However, coal use involves substantial technological complication due to the necessity of purifying the flue gas (the emission of sulfur and nitrogen oxides has to be minimized). Problems with dust catching and transportation and feeding of coal must also be solved. There is a wide range of clean coal technology options available. A fluidized bed combustor (FBC) system may be more suitable at a 200 MW power plants where a low quality fuel like coal can be used.[2] The advantages of this system include the ability to retain sulfur by adding limestone or another sorbent to the combustor, the inherently low NO_x emission due to reduced combustion temperature (800-900°C), and the wide range of fuels that can be accommodated. There are many facilities using this technology; the Nucla Demonstration Project employs an atmospheric circulating FBC shown in Figure 2.1.5.[3] The 110 MW combustor has a capacity of 42,000 kg/h and uses Salt Creek and Peabody coal (0.4-0.8% S) and Dochester coal (1.4-1.8% S) and limestone as a sorbent. The environmental results at 882°C included a 70% SO_2 removal with Ca:S ratio of 1.5:1 and 95% SO_2 removal with Ca:S=4:1. NO_x was less than 146 g/GJ, and combustion efficiency was 96.9-98.9%. The capital cost was approximately $1,123/net kW. As with any type of power plant, the flue-gas excessive temperature at 130-150°C produces a severe impact on the surroundings. There is currently not an adequate technology to reclaim such low-potential heat in the boiler system. All alumina refineries have an inexhaustible and unutilized byproduct available for the effective capture and coprocessing of power-plant gaseous, solid and thermal wastes - red mud. Annually, 50-55 million cubic meters of red mud containing 1.75 million tons of soluble and solid caustic are dumped throughout the world creating high environmental risk and adding to the losses of alumina and sodium hydroxide. [4]

108

Red Mud can be described as a multiple-mineral product. It does not dissolve in aluminate solution during the bauxite digestion; after washing and filtering it is directed to landfills. Main red mud constituents are hematite Fe_2O_3, anatase and/or rutile TiO_2, quartz SiO_2, tricalcium hydroaluminate $Ca_3Al_2(OH)_{12}$, and sodium hydroaluminosilicate $Na_2(AlSiO_4)_2.2H_2O$ (also called a desilication product /DSP/). Red mud may contain as much as 7-54% Fe_2O_3, 10-28% Al_2O_3, 4-24% TiO_2, 3-30% SiO_2, 5-15% CaO and 2-20% Na_2O, depending on mineralogical composition of the digested bauxite and additives (mainly lime) used in the Bayer process.[5] Also, some less common elements and rare metals, such as vanadium, yttrium, scandium and lanthanides, can be the part of red mud composition.[6] A high moisture content on the order of 25-45% and particle fineness (less than 50 microns of the average diameter) retard red mud transportation and application. On the other hand, storage cost affects the final alumina price significantly. Many methods have been suggested for red-mud processing.[5-7] Among them has been the use of red mud as a flocculant for water purification.[8,9] Red mud can be used to produced bricks [10] and polishable hard tiles [11] and employed as a rubber filler.[12] It is possible to produce a pigment by acidic treatment of a red mud followed by neutralization of a slurry obtained with carbonates.[7] The method for flue gases desulfurization using red mud was also presented.[13] Red mud can be used to slake lime treatment and to produce a blend applicable for rendering an incinerator's of the dioxin-containing flue gases harmless.[14] The oxides of iron, aluminum, and titanium that form red mud can be dissolved in mineral acids.[7] The majority of iron recovery at a level within 72-91% was obtained by leaching with 9N HCl. Alumina recovery in the HCl solution varies between 28% and 80% and titanium between 4% and 41%.[15] However, processes using acid cause caustic loss due to acidic neutralization into sodium salts. Iron, moreover, can be extracted from red mud by magnetic separation just as directly[16], and by drying, pelletizing with coal and lime, and reduction in a traveling circular-grate machine.[17] Red mud can be processed into alumina, caustic or soda ash, and calcium silicate by both alkaline hydrothermal and sintering methods. A hydrothermal alkaline process was applied to Kazakhstan's red mud. The following optimal parameters were determined and verified via the pilot-plant: temperature 280°C; NaOH concentration 520 g/L; molar ratios of $CaO:SiO_2$=1.3:1 and $Na_2O:Al_2O_3$=12:1; and extraction time 40 minutes. The yield of alumina was 88.8%.[18] Comalco's data from another bench-scale test were close to these parameters.[19] The Australian red mud was digested with caustic liquor in the presence of lime in an autoclave at 260-300°C to recover 70-80% Al_2O_3 and 95% Na_2O.

Conclusion. Shifting of alumina refinery from employment of both gaseous and liquid fuels to coal use opens the way to utilization of red mud and ash along with decrease of fuel cost. This advantage can be accomplished by integration of coal combustion and sintering red mud and ash as applied to the Bayer process used by an alumina refinery. Such conjunction permits to apply a fluidized bed combustor and reach low combustion-sintering temperatures 700-800°C, as well as clean and cool waste gases and also significantly reduce heat consumption. Furthermore, the alumina yield will increase and live caustic consumption decrease due to additional Al_2O_3 and Na_2O recovery from red mud and coal ash. The proposed integrated process represents main interest of alumina producers as a pattern for new generation of alumina refineries in 21st century which will be able to produce not only Al_2O_3 but also power, and byproducts, coincidentally with elimination of waste dumping.

References:

1. D.J. Donaldson, "Energy Savings in the Bayer Process," JOM, 09, 1981, p. 37-41

2. J.M. Topper, P.J.I. Cross, and S.H. Goldthorpe, "Clean Coal Technology for Power and Cogeneration," Fuel, 73 (7), 1994, p. 1056-1063.

3. "Clean Coal Technology Demonstration Program," brochure DOE ,Washington, DC, 1994.

4. N.C.R. Oeberg, "Red Mud and Sands Handling. New Thoughts on an Old Problem," Light Metals 1996, ed. Wayne Hale, Warrendale, PA: TMS, 1996, p. 67-63

5. V.A. Utkov, "Prospects for Development of Technology for Processing Red Muds in the USSR and Abroad," Moscow, Tsvetmetinformatsiya, 1983.

6. V.L. Rayzman, "Red Mud Revisited," Aluminum Today, 10(5), 1998, p. 64-68

7. L. Piga, F. Pochetti, and L. Stoppa, "Recovering Metals from Red Mud Generated During Alumina Production," JOM, 11, 1993, p. 54-59

8. B.K. Parekh and W.M. Golberger, "Utilization of Bayer Process Muds: Problems and Possibilities," Proc. of Mineral Waste Utilization Symposium, vol. 6, 1978, p. 123-132

9. G.Bayer,"Moglichkeiten zur Wirtschaftlichen Beseitigung von Rotschlammen," Erzmetal, 25 (9), 1972, p. 454

10. P.M. Prasad, "Metal Science - The Emerging Frontiers," ICMS-77, ed. T.R. Anantharaman et al., Calcutta, Indian Institute of Metals, 1979, p. 461.

11. C. Fernandez et al., "Use of Red Mud in Construction Materials," Light Metals 1996, Warrendale, PA: TMS, 1996, p. 99-106

12. A. Funke, H. Wetzel, and G. Buhler, "Rubber Fillers from Red Mud," DD patent 19,584 (Sept. 20, 1960)

13. J.V.S. Mani, R.C. Patel, and R.K. Saksena, "Utilization of Red Mud," Food Farm and Agr., 12 (10), 1980, p. 239-240

14. H. Bruhne, DE patent 3,436,085-A (March 10, 1986)

15. J. Pradhan, S.N. Das, S.B. Rao, and R.S. Thakur, "Characterization of Indian Red Muds and Recovery of their Metal Values," Light Metals 1996, Warrendale, PA: TMS, 1996, p. 87-92

16. M. Braithwait, GB patent 2,078,211-A (Jan. 06, 1982)

17. E. Guccione, "Red Mud, a Solid Waste Can Now Be Converted to High-Quality Steel,"

Eng. Min. J., 172 (9), 1971, p.136-138

18.L.P. Ni and V.L. Rayzman, "Combined Methods for Processing Low-Grade Aluminum-Containing Raw Material," Alma-Ata, Kazakhstan, Nauka, 1988.

19.P.J. Gresswell and D.J. Milne, "A Hydrothermal Process for Recovery of Soda and Alumina from Red Mud," Light Metals 1982, Warrendale, PA: TMS, 1982, p.227-238.

4.2.Creation of Power-Chemical-Metallurgical Complexes as the Way to Solve Both Raw Material and Fuel Problems in Aluminum Industry

Many developed and developing countries, like the United States, United Kingdom, Germany, Russia, Japan, and some others, don't dispose of industrial bauxite resources as a raw material for aluminum production. This is why the above-mentioned countries are forced to import bauxite, alumina, and/or metallic aluminum from abroad. In so doing, enormous expenses on raw material extraction, loading-unloading, and transportation are inevitable. At the same time, these countries are rich in deposits and reserves of coal. Often, mineral part of such solid fuel contains more than 20-25 per cent of alumina. Nevertheless, coal is combusted, or subjected to liquefaction or gasification, and the obtained alumina-bearing ash is either dumped or applied as not valuable raw material for road-building. Territorial joining raw materal sources with metallurgical plants would accomplish real economic and financial break-through in the aluminum industry, especially, taking into account the environmental improvement providing by such integration. Coal ash is among other potential sources for alumina production. Dumps of ash occupy vast expanses of our planet's territory. No less than 200 million tons of coal ash has been, e.g., annually formed in the U.S., Russia, and China /1/. Only little part of that waste is used for manufacturing road cover and construction materials /2/. The main coal ash body is irrevocably lost. At the same time, up to 10-15 million tons of aluminum per year could be worlwide produced only through utilization of coal ash. Considering that acid-resistant mullite $3Al_2O_3.2SiO_2$ is main alumina-bearing coal ash mineral, the most acceptable methods for processing coal-fired waste into alumina are both alkali-using processes: sintering and hydrothermal above-covered in this book (Chapter 2).

Technical-Economical Comparison of the Alkaline Methods Used for Processing Aluminosilicates. As an example for comparison of the processes a typical alumina-bearing coal ash composition can be considered. That sample contains SiO_2 53,1% and Al_2O_3 26,8% (molar ratio of $SiO_2:Al_2O_3=3,4:1$) /2/. Evaluation of material streams will be done below by correlation of simplified molar balance of the corresponding processes with no regard for calcium, iron and alkali percentages of ash, as well as excluding desilication of aluminate solutions procedure from consideration as not determining stage . The molar balance of the limestone sintering method (LSM) involving formation and leaching of calcium-aluminate sinter is shown in Figure 4.2.1.

112

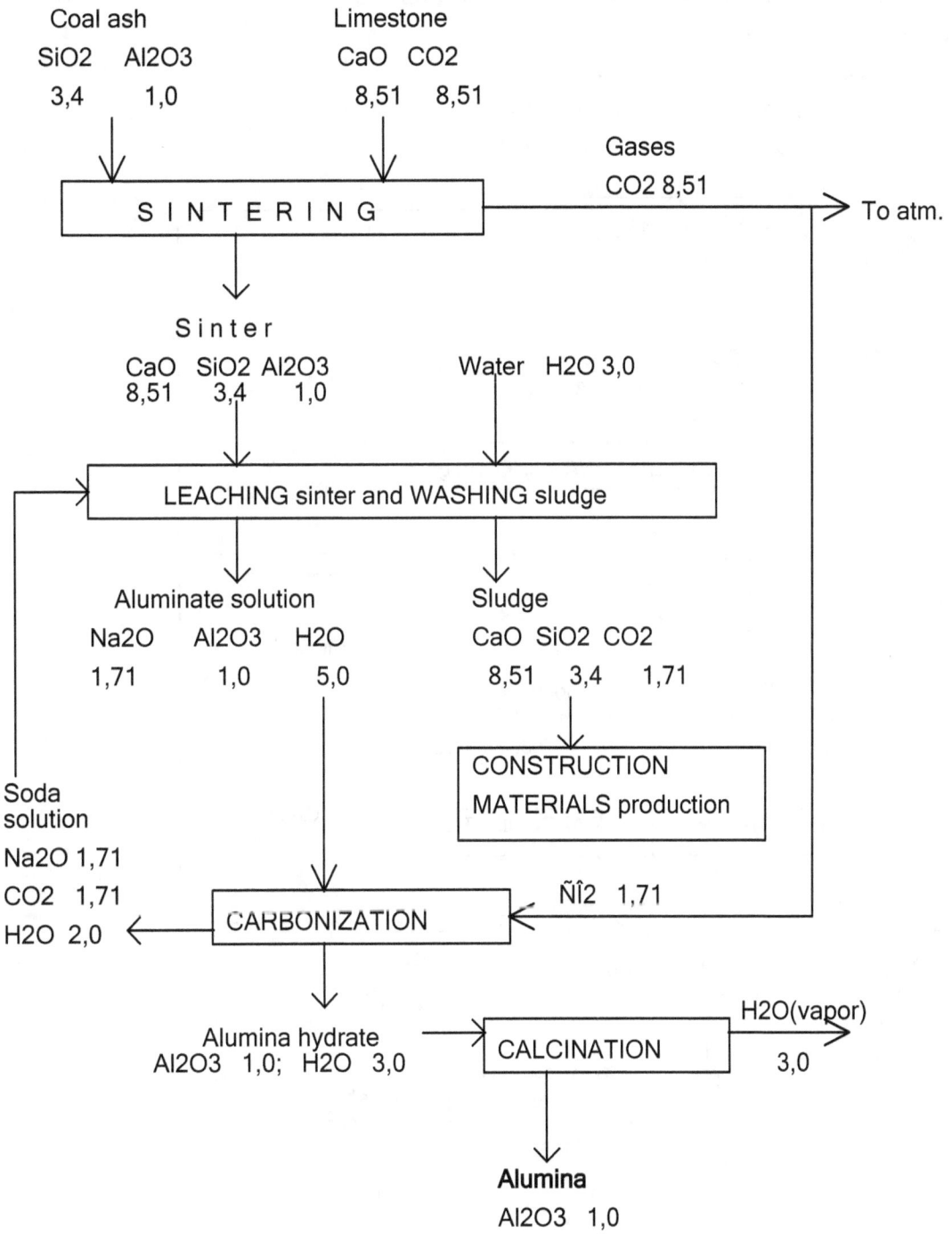

Figure 4.2.1. Molar balance of processing coal ash into alumina and byproducts by the method of sintering with limestone

As to limestone-soda sintering method (LSSM), which could be applicable for utilizing coal ash, or for procesings alumina concentrate produced from noncalcium ash by its chemical conditioning /3/, its molar balance is shown in Figure 4.2.2.

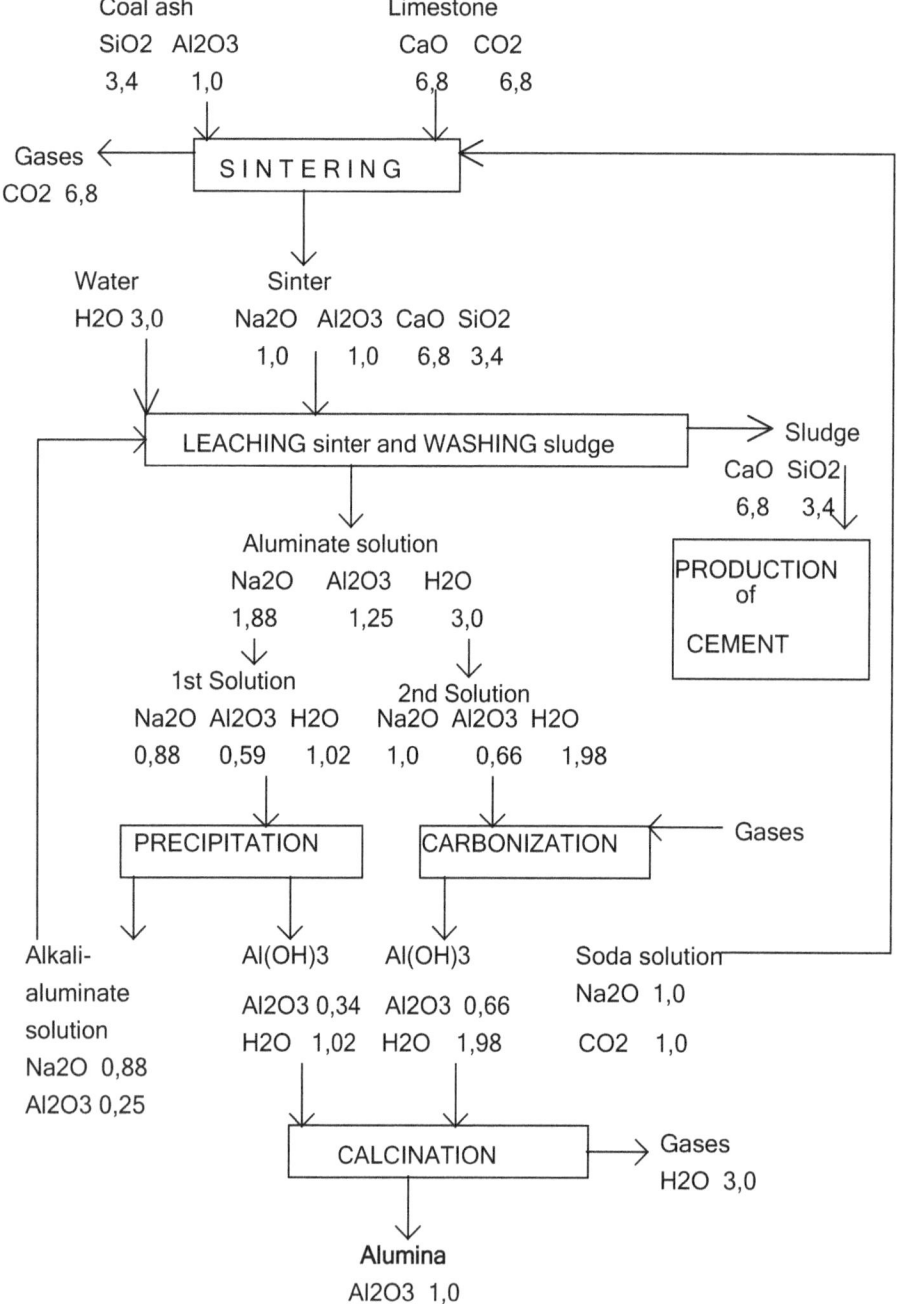

Figure 4.2.2. Molar balance of the ash-limestone-soda sintering method

The features of the hydrothermal alkali-lime method (HTALM) based on hydrothermal digestion of aluminoslicates at temperatures 250-300°C using strong ($\geq$350 Na_2O_{caust} g/L) alkali solution in the presence of lime are reflected in Figure 4.2.3.

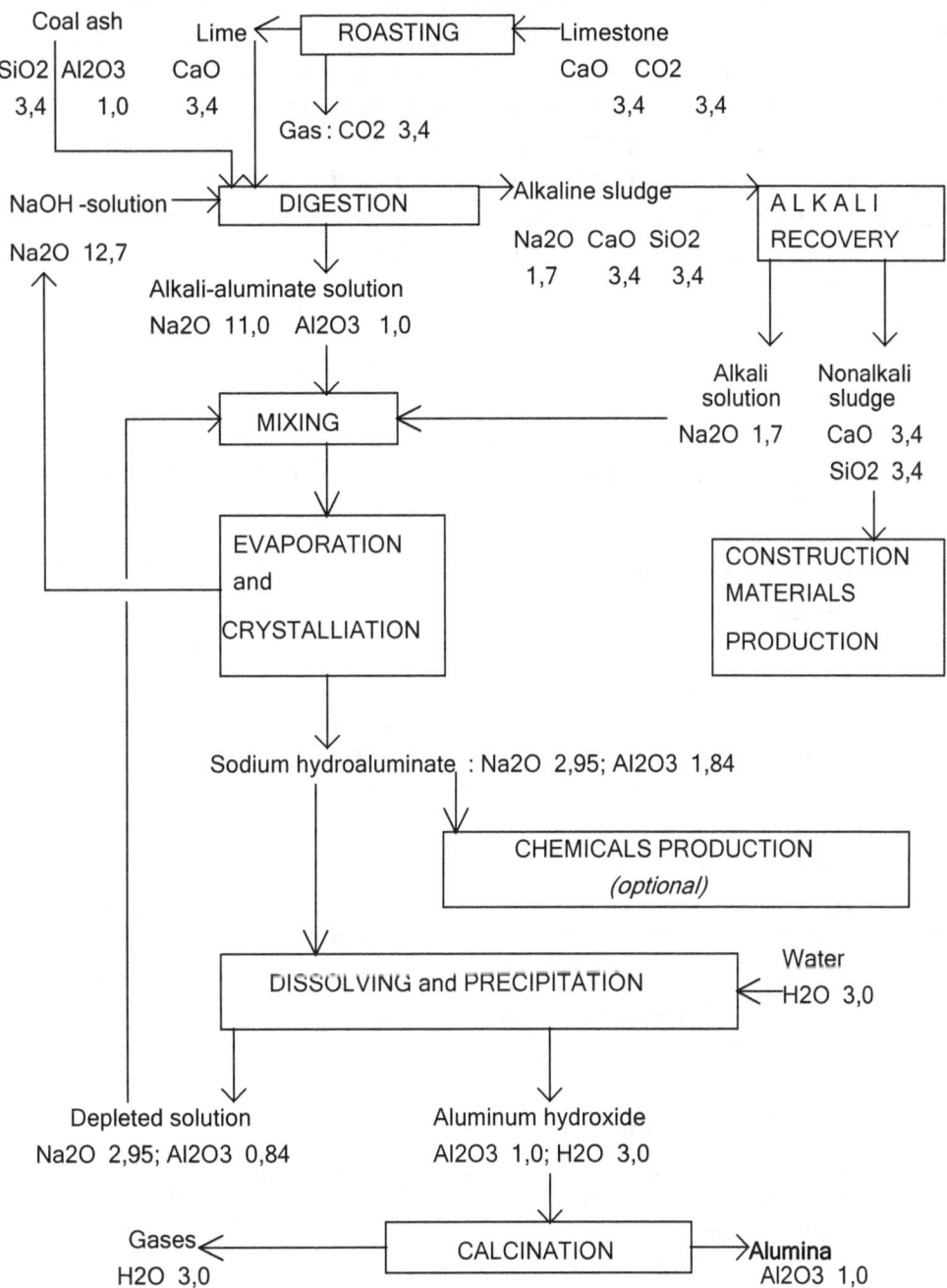

Figure 4.2.3. Molar balance of processing coal ash into alumina and byproducts by the hydrothermal method

The covered methods differ little from each other in magnitude of capital investment.[3] As to specific input of repeatedly renewable (nonregenerable) limestone ($CaCO_3$), its value for HTALM equal to 3.4 mol/mol Al_2O_3 that is 2-2.5

times less than those for both sintering processes - LSM (8.51 mol/mol Al_2O_3) and LSSM (6.8 mol/mol Al_2O_3). In other words, HTALM is more preferable on this parameter. Another one of the molar balance that characterizes process environmental protection is output of unutilized carbon dioxide formed in the course of both $CaCO_3$ and Na_2CO_3 calcination. In the case of HTALM, that factor equals 3,4 mol/mol Al_2O_3 whereas it rises for LSM to 6.8 mol/mol Al_2O_3) and for LSSM to 5.8 mol/mol Al_2O_3. Both hydrothermal and sintering methods were comparatively estimated in References /3,5,6/ as applied to processing aluminosilicate-containing feedstocks, namely, coal ash, nepheline and anorthosite. The obtained data are summarized in Table 4.2.1.

Table 4.2.1

Comparative technical-economical characteristics of the sintering and hydrothermal processes as applied to various types of aluminosilicate raw material (per 1 ton Al_2O_3)

Process:	SINTERING	H Y D R O T H E R M A L		
Feedstock :	Nepheline /5/	Nepheline /5/	Ash /6/	Anorthosite /3/
		Input		
Material(t:on)				
Feedstock	3.93	3.97	4.47	4.38
Limestone	7.92	3.64	5.65	4.09
Water	40	26	12.34	n.s
Power (kW-hr):	1050	600	1500	n.s
Fuel	mazut	coal	coal	coal
GJ:	71.6	47.7	19.25*	27.3
(USD):**	154.36	41.15	16.61	23.55
		Output of byproducts (ton)		
NaOH:		0.97		
(Na,K)$_2$CO$_3$:	1.08			
Sludge (on calculation to dry matter):				
	5.94	4.05	6.54	5.62

*Input of fuel on processing anorthosite by LSSM is equal to 65 GJ/t Al_2O_3 /3/, that is $101.98 using gas and $140.13 using crude oil.
**Fuel prices are taken from Reference Book [8].

According to conclusions of the above-mentioned works, expected fuel consumption as applied to nepheline comes to anorthosite 65 GJ/t Al_2O_3 for LSSM /3/, to 71.6 GJ/t Al_2O_3 for LSSM and 47.7 GJ/t Al_2O_3 for HTALM /5/, and to coal ash 19,25 GJ/t Al_2O_3 for HTALM /6/. Not only fuel consumption, but its type also determines that expense contribution. Hydrothermal method allows to make heat exchange between fuel gases and material blend through heating surface in contrast to direct gases-material contact in kiln as it takes place in the sintering procedure. For this reason, coal as the most cheap, readily available and wide-spread solid fuel can be used for heating "hydrothermal" alkali-lime aluminosilicate slurry on the schematic diagram as follows:

coal (shale) → fire-box→fuel gases→boiler→steam→heat exchangers. In this path, as well as in the worldwide Bayer process, the most perfect heat transfer and regeneration system is implemented due to application of multiple-stage slurry flashing and/or slurry-to-slurry heat exchange /7/. As to obtained calcium silicate sludge, the hydrothermal method causes decreasing its output around 1.5 times (counted to dry matter). Such decrease makes easier to overcome a difficulty in byproducts marketing. The reported comparison of the principal material and energetic parameters of both sintering and hydrothermal methods gives ground to consider HTALM as a prevailing technology for processing coal minerals into alumina. At the same time, despite the marked advantages, fuel consumption of HTALM is significantly (more than two times) higher than that of the Bayer method. Indeed, fuel input at the advanced alumina refineries processing low-silica bauxite is equal to 8 GJ per ton of alumina /9/. Such disadvantage can be equalized through integration of processing, on the one hand, fuel into steam and power and, on the other hand, alumina-bearing raw material including coal ash into alumina and aluminum /10/. In other words, we are dealing with creation of the unified power-chemical-metallurgical complex involving power plant, alumina refinery and aluminum smelter.

Integrated Technology as Applied to Different Types of Alumina-Bearing Raw Material. Use of coal ash as a source for alumina production is reasonable on condition that alumina-aluminum system would be built on to a power plant or to group of those being close together. A version of the integrated power-alumina-aluminum complex (IPAAC) flow-chart is schematically demonstrated in Figure 4.2.4.

117

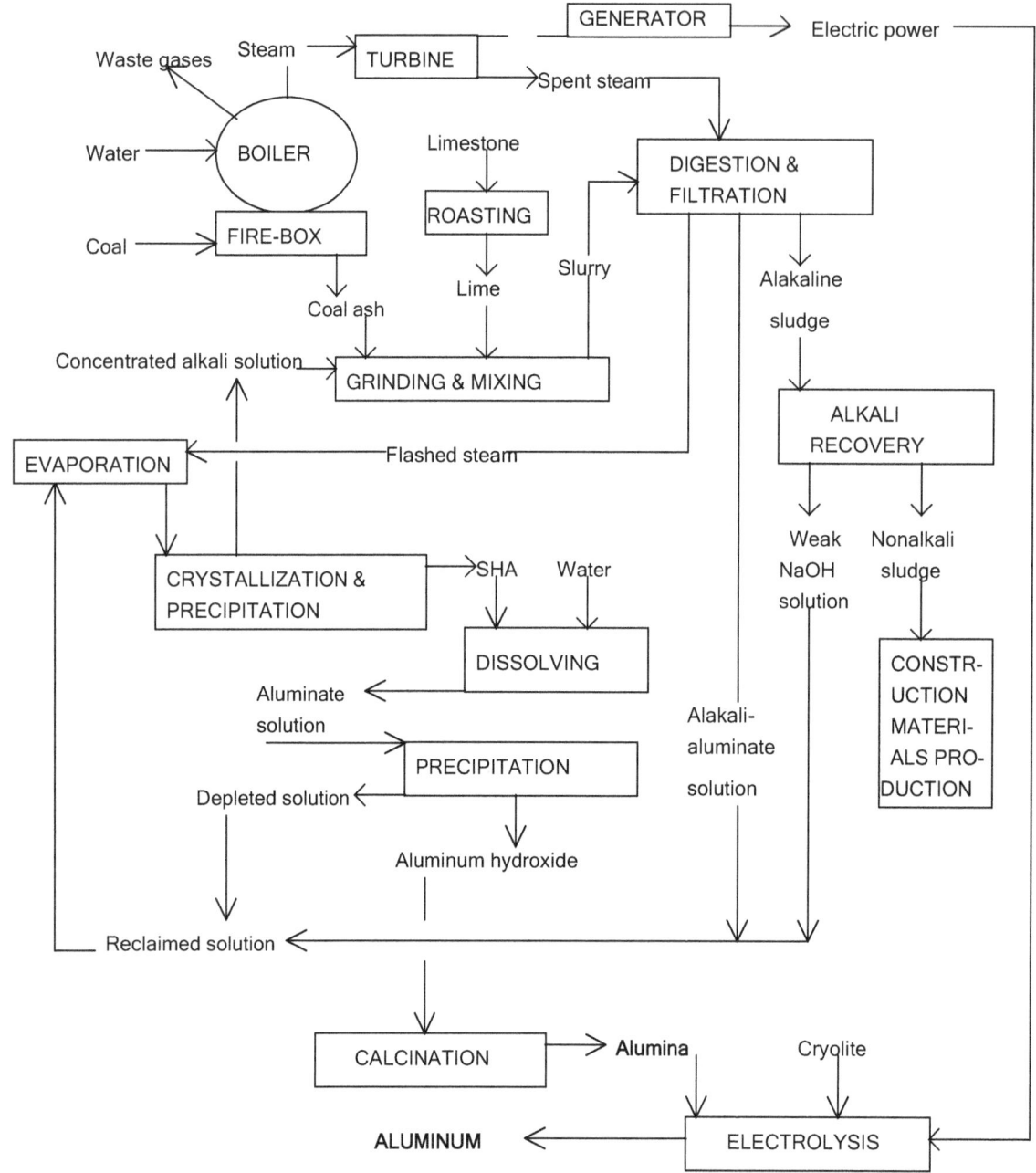

Figure 4.2.4. Conceptual flow chart of one of versions of the integrated process for manufacturing alumina, aluminum and electric power from coal

According to the shown lay-out, fuel gases obtained by coal combustion are used for steam production. The steam is derived from boiler to turbine bound with generator that works out electric power. Spent steam is headed after turbine to digestion unit of alumina refinery. The second combustion product - coal ash (in this case, we don't classified this material to fly and bottom ash) is ground and mixed with reclaimed concentrated alkali solution and lime obtained by calcination of limestone. The prepared ash-alkali-lime slurry is heated to 250-280°C at heat exchangers using the spent steam after turbine and held in digesters where both conversion of aluminum to solution and binding silicon into SCHS are completed. Cooling of digested slurry is carried out by means of multi-stage flashing; in so doing flashed vapor is used for evaporation of alkali-aluminate solution before SHA crystallization. The slurry is filtered or centrifuged and the separated slurry is washed and hydrolyzed for recovering NaOH as a weak solution (no more than 100 gNa_2O/L concentration). That NaOH liquor is joined with the filtered alkali-aluminate solution and depleted solution after decomposition (precipitation) procedure. Obtained nonalkali sludge is processed into construction materials. Joint solution is evaporated, then cooled, with the result that SHA is crystallized. Depleted concentrated alkali solution is reclaimed to the ash grinding procedure to prepare ash-alkali-lime slurry. The obtained SHA is water or process water dissolved and the obtained aluminate solution is subjected to precipitation procedure giving aluminum hydroxide which is calcined into alumina. This alumina oxide is headed to electrolysis of cryolite-alumina melt using electric power generated at the above-mentioned power plant. A drastic decrease in expenses for transportation and loading-discharging feedstocks, materials and fuel is evident advantage of IPAAC since, at the present time, both alumina refineries and thermal power plants generating electric power for aluminum smelters are, as a rule, well off each from other. Efficiency what is achieved by territorial approaching the above-mentioned enterprises is embodied in Table 4.2.2 wherein the Bayer method using bauxite, as well as LSSM, HTALM and IPAA process utilizing coal ash are compared on the principal varying expenses. As these expenses the following items are considered: cost of raw material, limestone and fuel, as well as expenditures in their loading-unloading and transportation. There was shown by Alcoa estimation that the capital investments to construction of alumina refinery operated with HTALM are not much distinguished from those of new standard plant operating abroad with the Bayer method /11/. On this reason, the corresponding expenses are not considered as varying ones. Bauxite price is taken equal $23,5 per ton on evidence of Reference /12/. Cost of expenditures in coal ash loading-unloading and transportation being also taken from the same reference is recalculated with consideration to ash alumina percentage and results $3.1 per ton of ash as applied to the sintering and hydrothermal processes. Since coal ash will be fed into IPAA complex directly from the power plant dust catching system, cost of expenditures in ash loading-unloading and transportation must be expelled from the comparative estimate. Price of limestone equal $8 per ton is taken from www.limestone-resources.com website. Input-output coefficients of the Bayer process are derived from Reference Book /13/.

119

Table 4.2.2

Comparison of economical characteristics for methods of alumina extraction from bauxite and coal ash on the varying expenses (per 1 ton of produced alumina)

Expense	Bayer process	Sintering process	HTALP	Integrated process
Materials:				
Feedstock	bauxite	ash	ash	ash
(ton)	2.05	3.83	3.83	3.83
(USD)	48.17	11.9	11.9	-
Limestone(ton)	0.125	10.8	4.94	4.94
(USD)	0.96	86.4	39.52	39.52
Fuel:	coal	oil	coal	coal
(GJ)	8.0	65.0	19.25	19.25
(USD)	6.9	101.98	16.61	16.61
Total				
(USD)	**56.03**	**200.28**	**68.03**	**56.13**

Data of Table 4.2.2 show that sum of expenditures at the considered variable items are as follows: $68.03 for HTALM that is significantly higher than the same sum for the Bayer process, whereas for IPAA method that characteristic is equal to $56.13, what is practically the same as the Bayer method's total expenditures. It is easy to verify that equalization of expenditures for both Bayer and IPAA methods is achieved due to decrease of loading-unloading and transportation expenditures. Creation and commercialization of IPAA complex will significantly allow to simplify and, consequently, to reduce the cost of supplying resource and power for aluminum industry works, as it is shown in Figures 4.2.5a and 4.2.5b.

There will not be required at IPAA complex processing alumina-bearing coal and shale to recover and grind bauxite, stockpile red mud at special equipped fields, build and operate ash dumps, consume expensive gas or oil and transport raw material and alumina . High-silicon bauxite also can be processed in the IPAA complex framework. A flow chart of such process version is shown in Figure 4.2.6 describing in-series manufacturing alumina from bauxite by the Bayer method and from the obtained red mud and coal ash by the hydrothermal technology. Initial bauxite and sodium hydroaluminate precipitated from the hydrothermal cycle solution are headed to the Bayer cycle.

120

5à.

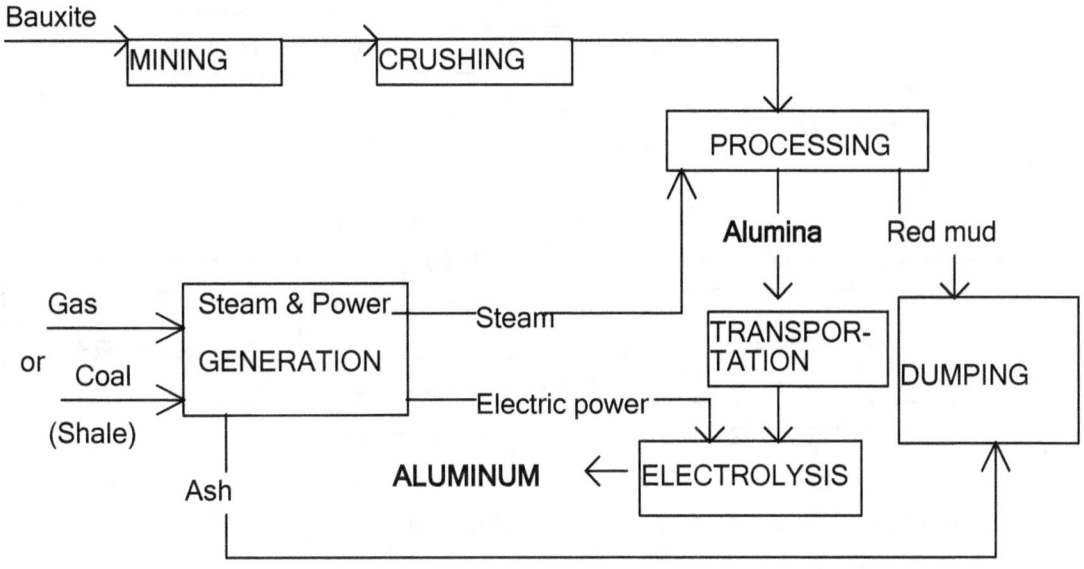

5b.

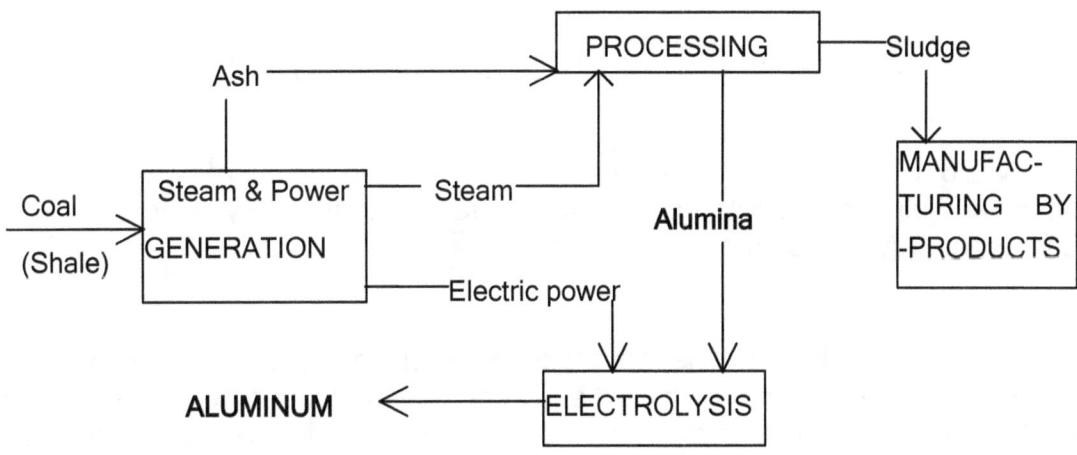

Figure 4.2.5. Flow chart for manufacturing steam, power, alumina and
  aluminum:
    a - by current separated technologies;
    b - by modified and simplified integrated power-chemical-metallurgical
      technology

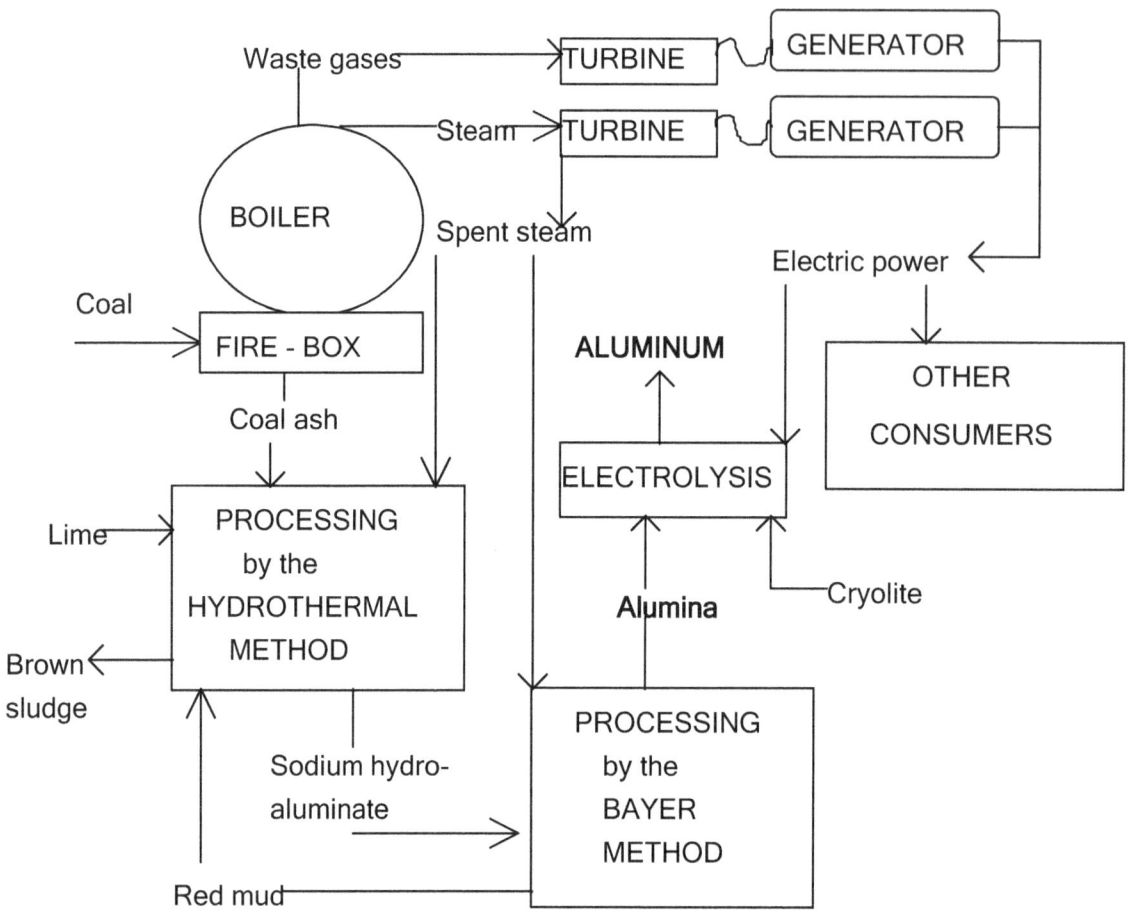

Figure 4.2.6. Conceptual flow chart of the combined production of power, alumina and aluminum from high-silicon bauxite and coal ash by using the in-series Bayer-hydrothermal method

Red mud is mixed with ash-lime intermediate product and processed by the hydrothermal method into brown mud and sodium hydroaluminate. Statement that the alternative Bayer-hydrothermal method is more attractive with application to high-silicon bauxite, is based on the technical-economical comparison of both competitive processes that has been accomplished before with respect to low-grade bauxite /5/. There was incorporated that compensation of alkali loss will be made by supplementation of nepheline concentrate. The results of the technical-economical comparison of both combined processes are given in Table 4.2.3.

Table 4.2.3

Input characteristics on processing low-grade bauxite with the combined alkali using processes (per 1 ton of produced alumina)

Input item	Bayer-Sintering method	Bayer-Hydrothermal method
Bauxite (16% of moisture)(ton)	2.142	2.222
Limestone (5%of moisture)(ton)	3.128	1.936
Nepheline concentrate(ton)	0.459	0.449
Steam(GJ)	10.50	15.88
Fuel (tons of reference fuel):	0.790	0.343
including: sintering	0.607	-
limestone calcination	0.029	0.188
$Al(OH)_3$ calcination	0.122	0.122
drying bauxite	0.032	0.033
Total heat input evaluated in tons of reference fuel	1.215	0.98
Electric power(kW-hr)	600	600

As it follows from the table, the Bayer-hydrothermal version is featured by lower heat consumption. Total fuel input for operation needs of this in-series method is 21% less than that of the Bayer-sintering process parameter. In so doing, input of high-quality hydrocarbon fuel in the Bayer-hydrothermal method 2.3 times lower than that in the Bayer-sintering version. Also, It is possible to combine coal combustion procedure with sintering the granulated blend that contains red mud, mineral coal constituents, limestone, and soda /14/. Distribution of material and thermal flows of that combined process is shown in Figure 4.2.7.

As it evident from the shown data, alumina yield as a result of extraction from bauxite, red mud and coal ash by IPAA method, involving in-series Bayer-sintering version is ranged up to 98,1%. Processing high-silicon bauxite in the IPAA complex is also of considerable industrial interest using the in-series technology involving the thermochemical beneficiation (TCB) of bauxite followed by the Bayer method for processing the obtained concentrate (see Chapter 1, part 1.2). The obtained concentrate is well suited to be processed into alumina by the traditional Bayer method. As to alkali-silicate solution being another TCB product, it is desilicated by treatment with lime or other calcium-containing reagent. As a result, silicon is precipitated as calcium hydrosilicate which is used for manufacturing construction materials. In such manner, production of , e.g., iron-free cement upon processing high-iron bauxite comes into possibility. Flow chart of IPAAC involving the TCB-Bayer consequent version is diagrammed in Figure 4.2.8.

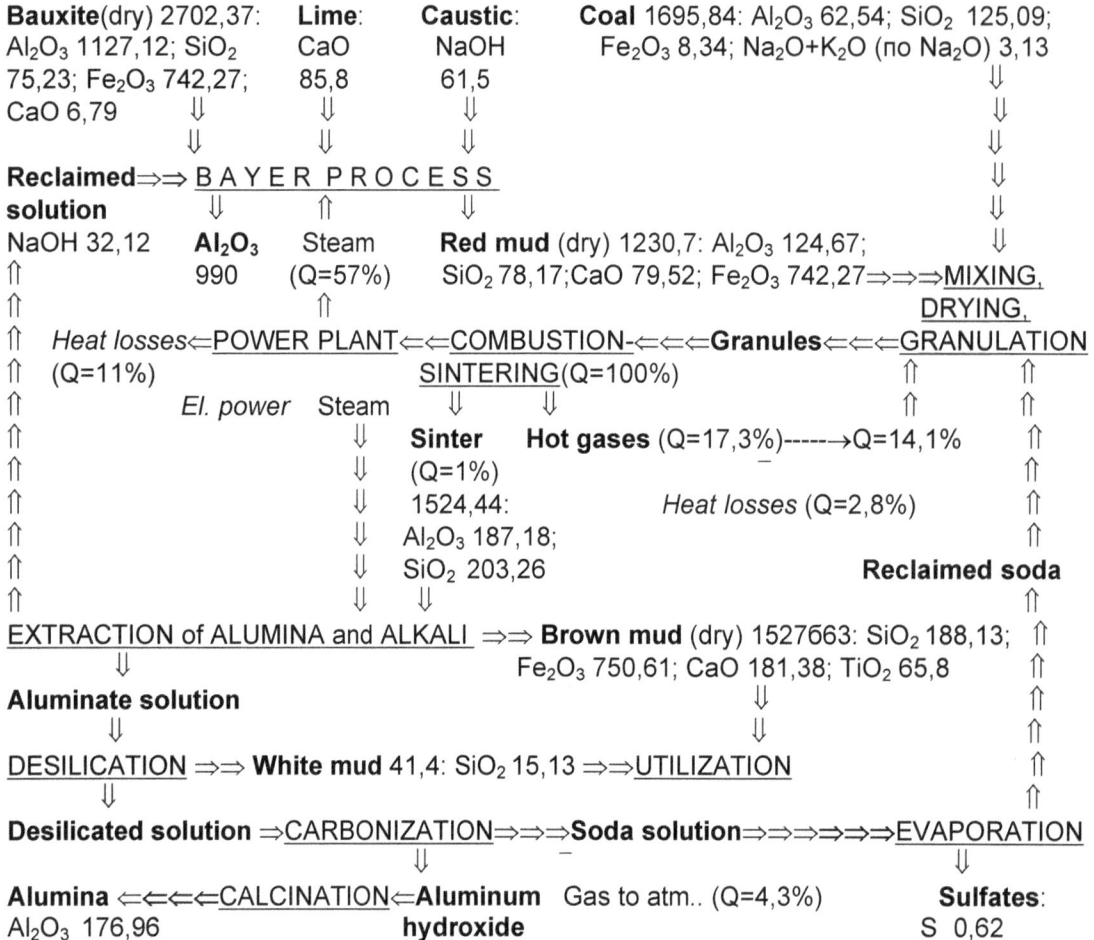

Bauxite(dry) 2702,37: **Lime**: **Caustic**: **Coal** 1695,84: Al_2O_3 62,54; SiO_2 125,09;
Al_2O_3 1127,12; SiO_2 CaO NaOH Fe_2O_3 8,34; Na_2O+K_2O (по Na_2O) 3,13
75,23; Fe_2O_3 742,27; 85,8 61,5   ⇓
CaO 6,79 ⇓ ⇓ ⇓ ⇓
  ⇓ ⇓ ⇓ ⇓

Reclaimed⇒⇒ B A Y E R P R O C E S S   ⇓
solution ⇓ ⇑ ⇓   ⇓

NaOH 32,12 **Al_2O_3** Steam **Red mud** (dry) 1230,7: Al_2O_3 124,67; ⇓
⇑ 990 (Q=57%) SiO_2 78,17;CaO 79,52; Fe_2O_3 742,27⇒⇒⇒**MIXING,**
⇑   ⇑   **DRYING,**
⇑ *Heat losses*⇐**POWER PLANT**⇐⇐**COMBUSTION-**⇐⇐⇐**Granules**⇐⇐⇐GRANULATION
⇑ (Q=11%)   SINTERING(Q=100%) ⇑ ⇑
⇑   *El. power* Steam ⇓ ⇓ ⇑ ⇑
⇑   ⇓ **Sinter** **Hot gases** (Q=17,3%)------→Q=14,1% ⇑
⇑   ⇓ (Q=1%) ⇑
⇑   ⇓ 1524,44: *Heat losses* (Q=2,8%) ⇑
⇑   ⇓ Al_2O_3 187,18; ⇑
⇑   ⇓ SiO_2 203,26 **Reclaimed soda**
⇑   ⇓ ⇓ ⇑

EXTRACTION of ALUMINA and ALKALI ⇒⇒ **Brown mud** (dry) 1527663: SiO_2 188,13; ⇑
⇓ Fe_2O_3 750,61; CaO 181,38; TiO_2 65,8 ⇑

Aluminate solution   ⇓ ⇑
⇓   ⇑

DESILICATION ⇒⇒ **White mud** 41,4: SiO_2 15,13 ⇒⇒UTILIZATION ⇑
⇓   ⇑

Desilicated solution ⇒CARBONIZATION⇒⇒⇒**Soda solution**⇒⇒⇒⇒⇒EVAPORATION
  ⇓   ⇓

Alumina ⇐⇐⇐⇐CALCINATION⇐**Aluminum** Gas to atm.. (Q=4,3%) **Sulfates**:
Al_2O_3 176,96 **hydroxide**   S 0,62

Figure 4.2.7.An example of both material (kg) and thermal (%) IPAAC balances, involving coal combustion, working out electric power and steam, and the in-series processing bauxite and red mud (together with coal ash) by the Bayer-Sintering method

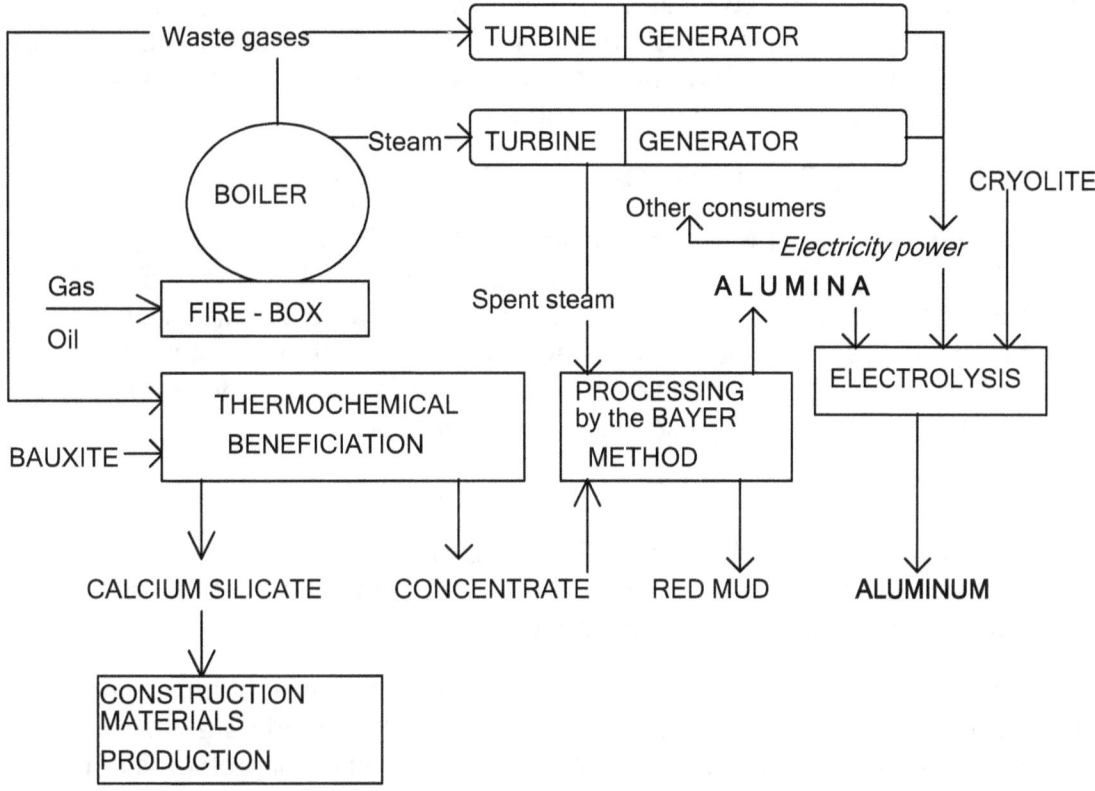

Figure 4.2.8. Principle flow chart of integrated production of electric power, alumina and aluminum from high-silicon bauxite by the in-series thermochemical beneficiation-Bayer process

There is more reasonable, in this modification, to combust ash-free (gas and mazut) or low-ash fuel in order that to prevent impuring alumina concentrate with ash.

Some Directions for Modification of the Integrated Technology. Territorial junction of both power and chemical-metallurgical works provides a way for combination of some productional operations, specifically, fuel combustion and limestone calcination /10/. It should be pointed out that temperature for carrying out the combined combustion-calcinations procedure must be within 900-1000°C to avoid lime overburning. Maintaining temperature in these required limits with simultaneous providing complete coal combustion is possible in fluidized bed furnaces using at the modern power plants in the U.S. /15/. Application of the fluidized-bed furnace makes possibility to ignite fuel-bearing blend containing up to 90% of non-combustible constituents /16/. This advantage allows to combine both coal combustion and limestone calcination in the common furnace with usage of produced lime for binding both ash constituents, namely, slicon into solid silicates and fly-sulphur into gypsum. The expected reduction of fuel consumption

125

will come to 9 GJ per ton of alumina /14/. Crisis of cement production caused by drastic decrease of consumers demands, can be overcome through soda leaching calcium hydrosilicate that is major constituent of the hydrothermal method solid waste - sludge. Being treated by soda solution with Na_2O_{carb} concentration 110 g/L at temperature 95°C and liquids to solids ratio 12,5:1, calcium hydrosilicate is decomposed into calcite $CaCO_3$ and soda-silicate solution /17/. Chemistry of that decomposition is described by the following reaction equation

$$CaH_2SiO_{4(s)} + Na_2CO_{3(aq)} = CaCO_{3(s)} + Na_2H_2SiO_{4(aq)}$$

Depending on number of the treatment steps, SiO_2 yield into liquor is increased from 53,8% (1 step) to 87,7% (4 steps). In such manner, either homogenic charge for producing cement or reclaimed CaCO3 can be prepared. Another obtained product of sludge decomposition, that is soda-silicate solution, is subjected to carbonization with waste gases after limestone calcination to produce commodity silica-gel and the reusable soda solution by the reaction

$$Na_2H_2SiO_{4(aq)} + CO_{2(gas)} = H_2SiO_{3(gel)} + Na_2CO_{3(aq)}$$

The cited examples don't exhaust the IPAAC features. According to solid fuel compositions and power plant locations, various versions of the integrated technology can be proposed and developed.

Conclusion. Combination of steam and power generation with alumina and aluminum production is, as it follows from the above-described data, the most effective solution of raw material, energetic and environmental problems of the corresponding industrial branches. In an effort for comprehensive revealing feasibilities and prospects of the proposed integrated technology, it is appropriate to pass into next stage, that is say, to establishment of the Demonstration project of IPAAC on the basis of some thermal power, alumina refinery, or aluminum smelter. Beside of the above-mentioned IPAAC advantages, its extremely important dignity is self-regulation or, in other words, independence from supplying raw material and fuel and, consequently, from energetic crises. Such autonomy will be useful for aluminum industry of developed and developing countries. In this connection, the Demonstration project implementation followed by its worldwide commercialization has to become key stage of aluminum industry development. It should be kept in mind that steam and electric power will be the IPAAC commodities along with alumina, aluminum, and byproducts. Such extension of products number, provide additional profitability of the created IPAAC.

References:

1. O.E.Manz, "Worldwide Production of Coal Ash and Utilization in Concrete and Other Products", FUEL (USA), Vol.76, No 8, 1997, p. 691-696.

2.V.L. Rayzman, S.A. Shcherban, and R.S. Dworkin, "Technology for Chemical-Metallurgical Coal Ash Utilization," Energy & Fuels, 1997, 11, p. 761-773. 3.K. Griotheim and Welch B., "Impact of

Alternative Processes for Aluminum Production on Energy Requirements," *JOURNAL OF METALS (USA), Vol. 33, No 9, 1981, p. 26-32.*

4. *V.L Rayzman and S.A Shcherban, "Recovering Alumina, Silica and Byproducts from Coal Ash Through the Use of Process for Silicon Pre-Extraction," Light Metals 1997, Warrendale, PA, TMS/AIME, 1997, p. 133-136.*

5. *V.S. Sazhin, "New Hydrochemical Methods for Complex Processing Aluminosilicates and High-Silicon Bauxites," Ìoscow, Metallurgiya, 1988, 213 p.*

6. *I. Podushkova and V. Novotna, "Extraction of Al2O3 from Nonbauxite Raw Material by Hydrometallurgical Method," Budapest, Bulletin of Scientific-Technical Council on Light Metals Production, SEV, 1979, 12, p. 157-165.*

7. *I.Z. Pevzner and V.L. Rayzman., "Autoclave Processes in Alumina Production," Ìoscow, Metallurgiya, 1983, 128 p.*

8. *Statistical Abstracts of the US Bureau of thr Census, Washington DC, 1996. 1024 p.*

9. *N.I. Eryomin, A.N. Naumchik, and V.G. Kazakov, "Processes and Apparatus in Alumina Production," Ìoscow, Metallurgiya, 1980, 360 p..*

10. *V.L. Rayzman, "Integrated Technology for Combustion of Aluminum-Containing Coal and the Recovery of Alumina and Byproducts from the Formed Ash," Light Metals 1996, Warrendale, PA, TMS/AIME, 1996, p. 151-157.*

11. *H.J. Hittner, "Hydrothermal Alkaline Process to Extract Àlumina from Anorthosite," TRAV. COM. INT. ETUDE BAUXITES, Alumine, Alum., Paris (France), 1981, 16, p.13-21.*

12. *W.B. Morrison and R.F. Nunn, "Future Bauxite Supplies to the Alumina Industry" (Poster), San Diego (CA, USA), Proceedings of TMS Annual Meeting, 1998.*

13. *L.B. Samaryanova and A.I. Lainer, "Technological Calculations in Alumina Production," Ìoscow, Metallurgiya, 1988, 256 p.*

14. *V.L Rayzman and I.K. Filipovich, "Integrating Coal Combustion and Red Mud Sintering at an Alumina Refinery," JOURNAL OF METALS (USA), 1999, 51(8), p. 16-18*

15. *U.M. Topper, P.J.I.Cross, and S.H. Goldthorpe, "Clean Coal Technology for Power and Cogeneration," FUEL, 1994, 3(7), p. 1056-1063.*

16. *B.S. Belosel'skii and V.I. Baryshev, "Low-Grade Energetic Fuels: Peculiarities of Preparation and Combustion," Ìoscow, Energoatomizdat, 1989, 136 p.*

17. *V.L. Rayzman, L.P.Ni, V.Ya. Abramov, and J.B. Rosen, "Study of Soda Decomposition Procedure as Applied to Sludge Obtained with the Hydrochemical Method," Nonferrous Metals, 1988, 6, p. 55-57.*

4.3. Thermochemical Evaluation of Industrial Methods for Processing Alumina-Bearing Raw Materials and Waste. *

The diversity of the range of ores and raw materials containing aluminum, causes the creation of a large number of technologies for manufacturing alumina and associated materials. The choice of technology depends on a rational cost-effeciency of a particular method

The effeciency of any method is determined by three factors: technological one f_1, factor in organizational and technical level of production f_2, and conjunctive factor f_3. The latter indicator depends on the availability of the material basis of production (labor, fuel, raw materials, reagents, transport, etc.). Factor f_2 characterizes the technical level of the equipment and process control facilities, safety devices to protect personnel and the environment from harmful emissions, the degree of heat recovery, and the entity of production.

In the absence of the impact of market fluctuations and technical shortcomings of production, that is, under idealized conditions, the economic efficiency E of the method is uniquely determined by a factor f_1.

Technology for producing alumina consists of a set of processes described by equations of the corresponding chemical reactions. Each of the reaction is characterized by the standard entropy change ΔS^0_{298}, calculated according to the law of Hess. Since the difference in the values of the entropy of substances indicates the difference in their nature / 1 /, the total standard entropy change of complex of chemical reactions is a quantitative characteristic of the overall process. The lower the total value ΔS^0_{298}, i.e. the higher the level of system ordering, the higher the objective efficiency of process.

In this regard, the technological factor f_1 can be determined from the expression
$$f_1 = \varepsilon \: / \: \Sigma \Delta S^0_{298} \qquad (1)$$

where ε - chemical extraction of useful component in fractions of a unit,

$\Sigma \Delta S^0_{298}$ -summary standard entropy change of the process chemical reactions in $kJ \cdot mol^{-1} \cdot K^{-1}$

Below are the results of calculations of the parameter f_1 for the most well-known technologies of processing aluminum-bearing raw materials such as boehmite, kaolinite, nepheline and albite. In the calculations, the following simplifications are accepted:

1) potassium in chemical interactions are not involved;

* - Rayzman, V.L, "Komleksn. Ispol'z. Miner. Syr'ya", 1985, 2, p. 51-55.

128

2) sodium hydroaluminosilicates resulting from alkali treatment of minerals and desilication of aluminate solutions have the chemical formula $Na_2(AlSiO_4)_2 \cdot H_2O$;

3) the product of the heat treatment of aluminum hydroxide is α-alumina.

Calculations are made with respect to 1 mole of Al_2O_3, so the chemical formula of solid reaction participants are presented for convenience in the form of oxides. Value of the entropy of interacting substances taken from handbooks / 2, 3 /. Composition aluminate ions at various process stages adopted according / 8 /, whereby in dilute aluminate solutions with concentrations up to 150 g / l Al_2O_3 $Al(OH)_4^-$ aluminate-ion form dominates, but in concentrated solutions (> 150 g / l Al_2O_3) the aluminate ions by the formula $Al_2O(OH)_6^{2-}$ prevail. Consider the calculation of the technological factor of most common technology for the processing boehmite $Al_2O_3 \cdot H_2O$, so-called Bayer process. It consists of the following chemical reactions:

1) Leaching
$$Al_2O_3 \cdot H_2O_{(s)} + 2OH^-_{(aq)} + H_2O_{(l)} = Al_2O(OH)_6^{2-}{}_{(aq)}$$

2) Dilution of the aluminate solution
$$Al_2O(OH)_6^{2-}{}_{(aq)} + H_2O_{(l)} = 2\,Al(OH)_4^-{}_{(aq)}$$

3) Precipitation
$$2Al(OH)_4^-{}_{(aq)} = 2Al(OH)_{3(s)} + 2OH^-_{(aq)}$$

4) Calcination
$$2Al(OH)_{3(s)} = \alpha\text{-}Al_2O_{3(s)} + 3H_2O_{(g)}$$

The overall equation of the process:

$$Al_2O_3 \cdot H_2O_{(s)} + 2H_2O_{(l)} = \alpha\text{-}Al_2O_{3(s)} + 3H_2O_{(g)}$$

According to the process as a whole:

$$\varepsilon = 1;\ \Sigma\Delta S^0_{298} = 0.381\ KJ \cdot mol^{-1} \cdot K^{-1};\ f_1 = 2.63\ mol \cdot K \cdot kJ^{-1}$$

Here are some general equation of a number of other processes. Processing kaolinite by the consecutive Bayer-hydrochemical method:

$$Al_2O_3 \cdot 2SiO_2 \cdot 2H_2O_{(s)} + 2.1222CaCO_{3(s)} + 0.097Na^+_{(aq)} + 0.1069OH^-_{(aq)} +$$
$$2.9532H_2O_{(l)} = 0.873(\alpha\text{-}Al_2O_{3(s)}) + 0.0485Na_2(AlSiO_4)_2 \cdot H_2O_{(s)} + 2.1222CO_{2(g)} +$$
$$1.8867CaH_2SiO_{4(s)} + 0.0785(3CaO \cdot Al_2O_3 \cdot 5.7H_2O \cdot 0.15\,SiO_2)_{(s)} +$$
$$0.0045H_2SiO_4^{2-}{}_{(aq)} + 2.6199H_2O_{(g)}$$

In general:

$\varepsilon = 0.873$; $\Sigma \Delta S^0_{298} = 0.659$ KJ·mol^{-1}·K^{-1}; $f_1 = 1.32$ mol·K·kJ^{-1}

Processing nepheline on alumina and sodium carbonate by the sintering method:

$Na_2O \cdot Al_2O_3 \cdot 2SiO_{2(s)} + 4.0759CaCO_{3(s)} + 4.1608H_2O_{(l)} + 0.0025OH^-_{(aq)} =$
$0.8922Al_2O_{3(s)} + 1.83Ca_2SiO_{4(s)} + 3.1583CO_{2(q)} + 2.6766H_2O_{(q)} + 0.34Ca(OH)_{2(s)}$
$+ 0.0824Na_2(AlSiO_4)_2 \cdot H_2O_{(s)} + 0.014H_2SiO_4^{2-}_{(aq)} +$
$0.0253(3CaO \cdot Al_2O_3 \cdot 5.7H_2O \cdot 0.15SiO_2)_{(s)} + 0.9176Na_2CO_3 \cdot H_2O_{(s)}$

$\varepsilon = 0.8922$; $\Sigma \Delta S^0_{298} = 0.75$ KJ·mol^{-1}·K^{-1}; $f_1 = 1.18$ mol·K· kJ^{-1}

Processing nepheline on alumina and sodium hydroxide by the hydrochemical method:

$Na_2O \cdot Al_2O_3 \cdot 2SiO_{2(s)} + 2.1222CaCO_{3(s)} + 5.9532H_2O_{(l)} = 0.8733Al_2O_{3(s)} +$
$2.1222CO_{2(q)} + 1.903Na^+_{(aq)} + 1.8939OH^-_{(aq)} +$
$0.0785(3CaO \cdot Al_2O_3 \cdot 5.7H_2O \cdot 0.15SiO_2)_{(s)} + 0.0045 H_2SiO_4^{2-}_{(aq)} +$
$1.8867CaH_2SiO_{4(s)} + 0.0485Na_2(AlSiO_4)_2 \cdot H_2O_{(s)} + 2.6199H_2O_{(q)}$

$\varepsilon = 0.8733$; $\Sigma \Delta S^0_{298} = 0.372$ KJ·mol^{-1}·K^{-1}; $f_1 = 1.72$ mol·K· kJ^{-1}

Processing kaolinite by the nitric acid method:

$Al_2O_3 \cdot 2SiO_2 \cdot 2H_2O_{(s)} + 13H_2O_{(l)} = Al_2O_{3(s)} + 15H2O_{(q)}$

$\varepsilon = 1$; $\Sigma \Delta S^0_{298} = 1.864$ KJ·mol^{-1}·K^{-1}; $f_1 = 0.54$ mol·K· kJ^{-1}

The results of the thermochemical calculations are shown in Table 4.3.1.

Table 4.3.1.

Results of Calculation of Technological Factor f_1 for Different Methods of Processing Alumina-Bearing Minerals

Factor f_1, mol·K·kJ^{-1}	Boehmite	Kaolinite	Nepheline	Albite
2.63	Bayer method			
2.14		Hydrochemical method with sodium hydro-aluminate production		
1.72		The same process with alumina production		
1.32		Serial version of Bayer-Hydrochemical method		
1.31		Hydrochemical method		
1.18			Classical Sintering method	
1.03			Sintering method using high-alkali blend	
0.85				Serial version of Chemical Beneficiation-Hydrochemical method
0.82				Hydrochemical method
0.63		Sintering method		
0.60				Chemical Benefication-Sintering method
0.54		Nitric acidic method		

As to boehmite processing, the Bayer's method is the most technologically advanced way ($f_1 = 2.63$ mol·K· kJ^{-1}). With regard to kaolinite the serial version of Bayer-hydrochemistry is preferable ($f_1 = 1.32$ mol·K· kJ^{-1}), as well as the direct

hydrochemistry method ($f_1 = 1.31$ mol·K· kJ^{-1}). The lowest technological factors for kaolinite have both sintering ($f_1 = 0.63$ mol·K· kJ^{-1}) and acid ($f_1 = 0.54$ mol·K· kJ^{-1}) processes. The most promising method of decomposition of nepheline to obtain alumina should recognize hydrochemical technology ($f_1 = 1.72$ mol·K· kJ^{-1}), the least promising - a sintering method using high-alkali blend ($f_1 = 1.03$ mol·K· kJ^{-1}).

It seems reasonable to use the version for hydrochemical processing nepheline on sodium hydroaluminate. The implementation of this option ($f_1 = 2.14$ mol·K· kJ^{-1}) would successfully solve the problem of transmission of caustic alkalies of nepheline to bauxite treatment circle, that is the Bayer process. Hydrochemical method has a leading position among other technologies. During the processing of albite, for example, preferred the serial version of the chemical beneficiation-hydrochemistry technology ($f_1 = 0.85$ mol·K· kJ^{-1}) and direct hydrochemical processing (f1 = 0.82 mol·K· kJ^{-1}) compared with the serial method of chemical enrichment-sintering ($f_1 = 0.6$ mol·K· kJ^{-1}).

The proposed factor as a criterion for a rapid evoluation of the efficiency of the processing various minerals and waste is less subjective measure than the cost of the end products, and can be used for determining the prospects of a particular technology.

References

1.Karapet'yants, M.Kh .; Chemical Thermodynamics. Moscow: 1975, 584 p.
2. Karapet'yants, M.Kh., Karapet'yants, M.L.; The Basic Thermodynamic Constants of Organic and Inorganic Materials: Reference Book. Moscow: Khimiya, 1968, 472 p.
3.Ryabin, V.A., Ostroumov, M.A., Svit, T.F .; Thermodynamic Properties of Materials: Reference Book. Leningrad: Khimiya, 1977, 392 p.
4.Burkov, K.A, Sizyakov, V.M, Myund, V.M, et al; Zhurnal Prikl. Khimii, 1979, 52, # 1, p. 53-57.
5.Lainer, A.I. Alumina production. Moscow: Metallurgiya, 1961, 620 p.
6.Pevzner, I.Z., Rayzman, V.L .; Autoclave Processes in Alumina Production. Moscow: Metallurgiya, 1983, 128 p.
7.Abramov, V.Ya., Sazhin, V.S, Shmorgunenko, N.S, Medvedev, V.V; Zhurnal "Tsvetnye Metally", 1983 # 11, p. 29-31.

ATTACHMENT

PROPOSALS ON A PROJECT "NEW TECHNOLOGY FOR INTEGRATING PRODUCTION OF POWER, CHEMICALS, AND ALUMINUM"

1.THE ESSENTIALS of the PROJECT

1.1. The Project Name: "NEW TECHNOLOGY FOR INTEGRATING PRODUCTION OF POWER, CHEMICALS, AND ALUMINUM"

1.2. Scientific-Technical Avenue which that Project Related to

Production of power, alumina, and aluminum from coal that minerals contain 20 and more per cent of alumina.

1.3.Main Definitions (Key Words)

Ash, coal, alumina, aluminum, electric power, steam, hydrochemical, hydrothermal, alkaline, lime, combustion, sodium hydroaluminate.

1.4.Actuality of the Project

The project is directed for solving problem of fuel and alumina-bearing raw material providing the countries, those don't posses the sufficient commercial reserves of bauxite, gas and oil and which economics depends on supplying these resources from abroad.

1.5. Operation Principle, Design, Achieved Scientific-Technical (Technical, Structural, Technological) Results

The project is the first proposed process combining three technologies in the frame of the united industrial complex. The combined technologies are the following works:

1)processing alumina-bearing coal into steam and electric power, or gaseous and liquid fuel ;

2)utilization of the formed coal ash by application of the hydrothermal (hydro-chemical) lime-alkali method producing alumina and other chemicals;

3)electrolysis of cryolite-alumina melt producing metallic aluminum.

All parts of the integrated process either are industrial used: generation of steam and electric power, production of sodium aluminate, decomposition of aluminate solution, calcinations of alumina hydrate, smelting aluminum; or were pilot-scale tested at the plants of VAMI (Russia) /V.S. Sazhin, "New Hydrochemical Methods of Completed Processing Aluminosilicates and High-Silicon Bauxite ", Moscow: Metallurgiya, 1988, 213 p./, Alcoa (USA)/H.J. Hittner, "Hydrothermal Alkaline Process to Extract Alumina from

133

Anorthosite," TRAV. COM.INT. ETUDE BAUXITES, ALUMINE, ALUM., Paris, France, v. 16, 1981, pp.13-21/ and ZSNP (Slovakia)/I. Podushkova and V. Novotna , "Production of Al_2O_3 from Nonbauxite Raw Material by the Hydrometallurgical Method", Bulletin of Scientific-Technical Committee on Light Metals Production, SEV, Budapest, No 12, 1979, pp. 157-165/: lime-alkali autoclave digestion of aluminosilicates, separation of autoclave slurry into alkali-aluminate solution (AAS) and alkaline sludge, recovery of alkali from the obtained sludge, and AAS evaporation.

1.6. Scientific, Applied, and Consumers Innovations of the Project Scientific innovation

Scientific innovation:Thermochemical and thermodynamical basics of technological and thermal integrations of steam and electric power generation, production of alumina, aluminate and silicate products and smelting aluminum.

Applied innovation: Creation and introduction of the united self-supported power-chemistry-metallurgy complex, characterized by minimal transportation, energetic, and operational expenditures and providing environmental protection of the surroundings.

Consumers novelty: Providing producers of power, alumina, aluminum and chemical byproducts with local fuel and raw material resources, as well as elimination of local producers dependence from foreign raw material providers .

2. FIELDS of APPLICATION and POTENTIAL MARKET EVALUATION

2.1. Possible fields of application. Expected scientific, technical, economical, and social effects resulted by commercialization of the project

Possible fields of application: combination of technologies for processing alumia-bearing coal either by combustion at power plant, or gasification, or liquefication coupled with utilization of the obtained solid remainder (coal ash) into alumina and byproducts by the hydrothermal method.

Scientific benefit of the project usage :

-Extension and deepening conceptions about mechanism of phase conversions and equilibrium states in the following systems: $Na_2O-Al_2O_3-CaO-SiO_2-H_2O$, $Na_2O-Al_2O_3-SiO_2-H_2O$, $Al_2O_3-CaO-SiO_2$, etc.

-Study of kinetics of simultaneously running thermal interactions of ash minerals conversion and limestone calcinations.

Technical efficiency: Integration of coal combustion and limestone calcination in fluidized bed; Soda treatment of no alkali sludge producing homogenized charge for cement production; Development of systems for pneumo- and hydro-transportation of dry and wet materials between the IPAAC units; Modification of apparatus arrangement of the hydrothermal method applying the newest

134

advances in the field of designing facilities of energetic, chemistry and metallurgy.

Economical efficiency: Savings of IPAAC commercialization as compared with the industrial analog that is the sintering method is more than USD 140 per ton of the produced alumina at the cost of just as significant decreasing materials and fuel transportation, loading, and shipping expenditures, so due to substituting high expensive hydrocarbon (gas, oil) fuel by cheaper solid those (coal, shale)

/V.L. Rayzman, V.M. Sizyakov, I.Z. Pevzner, V.I. Pevzner, and N.N. Sabadash , "Creation of Power-Chemistry-Metallurgy Complexes as a Way fot Solving Raw Material and Fuel Problems of Russia's Aluminum Industry", Tsvetnye Metally, No 2, 2003, pp.47-55/ (see Table 1).

Social effect: Creation of new working places.

Environmental efficiency:

-liquidation of dumps for storing both toxic wastes - coal ash and red mud ; - lowering amount of exhausted gaseous carbon dioxide 1,5-2 times in comparison with the industrial sintering method applied in Russia, China, and Kazakhstan.

2.2. Approximate estimation of the world market that is suitable for the proposed project

The total amount of coal ash produced only in three largest coal consumed countries - USA, Russia, and China is 215 million tons per year /Î.À. Manz, "Worldwide Production of Coal Ash and Utilization in Concrete and Other Product," FUEL, vol.76,No 8, 1997, pp.691-696/. About a quarter of that amount is coal ash containing, in average, 25% Al2O3 /M.Ya. Shpirt, V.R. Kler, and I.Z. Persikov , "Inorganic Constituents of Solid Fuel ",Moscow: Khimiya, 1990.-240 p./. 18 million tons of alumina could be produced processing such coal ash amount, yearly, that is around 40% of the world alumina production.

2.3.Competitors of the proposed integrated technology

The competitors are, on the one hand, all power and chemical companies and enterprises processing alumina-bearing coal and discarding ash with high alumina percentage. On the other hand, the competitors are all alumina refineries those, as is generally known, don't process ash as alumina-bearing raw material. No attempts to integrate power and alumina production using coal as both fuel and feedstock were yet industrially made.

3.DISADVANTAGES OF ANALOGES (PROTOTYPES) and ADVANTAGES OF THE PROPOSED TECHNOLOGY

By now, a large body of varied methods for production of alumina from minerals of coal have been proposed and investigated. The late 1960s oil crisis was a stimulus for these methods development. It was suggested that extraction and usage of coal as an alternative fuel will be increased. The developed methods for coal ash processing into alumina can be classified as follows /V.L. Rayzman, S.A. Shcherban, and R.S. Dworkin, "Technology for Chemical-Metallurgical Coal Ash Utilization", ENERGY&FUELS, 011 (004), 1997, pp. 761-773/:

1)Acidic extraction of aluminum and other metals contained in coal;

2)Method for ash sintering (MAS);

3)Combined coal combustion and ash sintering;

4)Hydrothermal method.

High temperatures of the conventional coal combustion procedure (more than 1000°C) cause formation of acid-resisting mineral of mullite. Therefore, using alkali as an alumina leaching reagent is more preferable compared with acid by coal ash processing. The known method for ash sintering requires to use high temperatures (1200-1500°C) and it is carried out under direct contact of the burned fuel with the formed ash and both carbonate reagents -soda and limestone. The main argument against MAS commercialization is necessity to apply expensive hydrocarbon fuel like oil, mazut, and gas. The combined method for alumina production through sintering mixture of alumina-bearing raw material with lime or limestone at temperature 1500°C followed by the obtained sinter leaching and carbonization of the produced aluminate solution /N.I. Eryomin, S.A. Tager, V.N. Kostin, et al. "Method for Alumina Production ", SU 371774, Oct. 25, 1977/.

However, conditions of the boiler fire-box operation doesn't provide sufficiently long time of contact between the blend constituents for completing sinter formation. In so doing, the formed chemicals are not alkali aluminate and calcium silicate, as it was expected, but they are defined as "undersinter", from which aluminum cannot be extracted. As to the hydrothermal method, that process is based on lime-alkali digestion of coal ash at temperatures 250-280°C. As a result, formation of Alkali-Aluminate Solution (AAS) and solid Sodium-Calcium Hydrosilicate (SCHS) takes place. The solids being so-called alkaline sludge is subjected to hydrolysis to liberate alkali as the reclaimed NaOH-solution and produce no alkali sludge containing Monocalcium Hydrosilicate (MCHS). Alkali solution is evaporated then cooled to 45-50°C and solid Sodium Hydroaluminate (SHA) is crystallized. The obtained crystalls is water dissolved giving aluminate solution that is subjected to precipitation producing aluminum hydroxide. Finally, $Al(OH)_3$ is calcined into alumina Al_2O_3. Comparison of principal material and energetic characteristics of the above-mentioned processes gives grounds to conclusion that the hydrothermal method is more preferable as applied to utilization of coal ash alumina-bearing minerals /V.L. Rayzman, P. Rerkjirasawad, and A. Aturin, "Thailand's Lignite Ash as a Raw Material for Manufacturing Alumina and Byproducts by Alkaline Hydrothermal Method", SURANAREE JOURNAL OF SCIENCE AND TECHNOLOGY, vol. 8, No 1-2, 2001, pp. 83-898/. Nevertheless, in spite of the certain dignities, the hydrothermal process is characterized with fuel consumption that is significantly (2-3 times) higher than that of the worldwide Bayer method used for production alumina from bauxite. Input of consumed fuel for the Bayer process is within 7-8 GJ per ton of commodity alumina /N.I. Eryomin, A.N. Naumchik, and V.G. Kazakov,

"Processes and Facilities in Alumina Production", Iiscow, Metallurgiya, 1980, 360 p./. The noted disadvantage of the hydrothermal method can be largely overcome by employing the proposed new integrated technology */V.L. Rayzman, "Integrated Technology for Combustion of Aluminum-Containing Coal and the Recovery of Alumina and Byproducts from the Formed Ash," LIGHT METALS (TMS), Warrendale, PA, 1996, pp.151-157/*.

<u>The main advantages of the proposed integrated method in comparison with its analog.</u> The proposed integration considers creation of the united power-chemical-metallurgical complex involving power plant, alumina refinery, and aluminum smelter. In such complex, utilization of coal ash can be carried out by building on alumina-aluminum enterprise to the operating thermal power plant (or to group of thermal power plants). Fuel gases, produced in the course of coal combustion, are applied for working out steam that heads from boiler to turbine, The turbine is connected with generator producing electric power. Spent steam after the turbine is heat carrier for coal ash digestion. That procedure is conducted by heating alkali-lime-ash slurry to 250-280°C followed by residing slurry at these temperatures in digesters. Then digested slurry is cooled by flashing and filtered into AAS and alkaline sludge. The obtained sludge is water washed and gives away its alkali as NaOH-solution. Then alkali-free sludge is processed into cement and other construction materials. Extracted NaOH-solution is mixed with AAS and depleted solution after $Al(OH)_3$ precipitation. The mixed solution is evaporated, then cooled and sodium hydroaluminate is precipitated. That solid intermediate is dissolved and obtained aluminate solution is subjected to $Al(OH)_3$ precipitation. Then obtained aluminum hydroxide is calcined into alumina that is processed into aluminum by electrolysis at metallurgical plant (smelter). Significant decreasing expenditures for transportation, shipping and loading raw material, reagents and fuel is obvious argument of efficiency of the proposed integrated technology since, at the present time, alumina refineries and power plants are far apart. Economical advantages achieved by territorial approaching all enterprises of IPAAC are reflected in Table 1. In that table the Bayer method (as applied to bauxite) is compared with the sintering, hydrothermal ind integrated processes (all those as applied to coal ash) using only varying expenditure items. These items is considered as transportation, shipping and loading expenditures, as well as cost of fuel. There has been found before */H.J. Hittner, "Hydrothermal Alkaline Process to Extract Alumina from Anorthosite," TRAV. COM.INT. ETUDE BAUXITES, ALUMINE ALUM., Paris, France, v. 16, 1981, pp.13-21/* that expenditures for construction and operation of both plants processing bauxite by the Bayer method, and aluminosilicates by the hydrothermal process are close to each other. All cost characteristics in Table 1 come in USD. Price of bauxite, as well as cost of materials transportation, shipping and loading are taken on evidence of Reference */W.B. Morrison and R.F. Nunn, "Future Bauxite Supplies to the Alumina Industry," Proceedings of TMS Annual Meeting, San Diego (CA, USA), 1998 (Poster)/*. In so doing, transportation, shipping and loading expenditures are recalculated for the sintering and hydrothermal methods taking into account lower ash alumina percentage. According to calculations cost of transportation, shipping and loading one ton of ash is equal to $ 3.1. In the case of IPAAC commercialization, coal ash feeds into processing directly from ash traps, therefore is not necessary to transport and load coal ash. Limestone price is taken

on evidence of the website www.limestone-resources.com. Input characteristics for the Bayer process are taken from Reference book /L.B. *Samaryanova and A.I. Lainer, "Technological Calculations in Alumina Production," Moscow: Metallurgiya, 1988, 256 p./*. As it follows from Table 1, expenditures on the varying items for the hydrothermal method ($ 68.03 per ton of alumina) are significantly higher than those for the Bayer process ($56.47/ton Al_2O_3), whereas the same cost parameter for the proposed integrated technology ($56.13/ton Al_2O_3) is practically equal to that for the Bayer method. It is easy to verify that such leveling cost parameters is caused, in first turn, by sharp decreasing materials transportation, shipping and loading expenditures. In addition to beneficial economic indexes, IPAAC is characterized by doubtless environmental advantages due to ash dump liquidation and decreasing amounts of the exhausted CO_2, since input of limestone is two times lower compared with the sintering process currently implemented for aluminosilicate (nepheline and red mud) utilization.

Table 1 Comparison of economical characteristics of the processes for production alumina from bauxite and coal ash on the varying expenditure items (per 1 ton of commodity alumina).

Item/Process	Bayer method	Sintering method	Hydrothermal method	Integrated method
Materials:				
Feedstock	bauxite	ash	ash	ash
(tons)	2,05	3,83	3,83	3,83
(USD)	48,17	11,9	11,9	–
Limestone				
(tons)	0,125	10,8	4,94	4,94
(USD)	0,96	86,4	39,52	39,52
Fuel	coal	mazut	coal	coal
(GJ)	8,0	65,0	19,25	19,25
(USD)	6,9	101,98	16,61	16,61
Total				
(USD):	56,03	200,28	68,03	56,13

4. THE PROBLEMS AND AVENUE OF DEVELOPMENT

4.1. Economical characteristics

Creation of the Demonstration Project on the base of one enterprise of the proposed complex, supposedly, the thermal power plant processing alumina-bearing coal should be the first stage of the integrated technology commercialization.

4.2. Bulk, time and cost of work

Total time of the preliminary investigations – two years. The cost is depending on the country economics level and local conditions.

List of measures intended for preparing to design and industrial mastering the integrated technology

1) Sampling and delivering representative consignments of coal and coal ash from the thermal power plant selected as an IPAAC enterprise

2) Refinement of coal combustion parameters obtaining sample of ash of optimal composition.

3) Carrying out the parallel control bench-scale tests using both ash samples - delivered from the thermal power plant and obtained by the same plant coal bench-scale combustion obtaining comparative technological data on the following units :

-ash digestion,

-filtration of the digested slurry,

-recovering alkali from the washed alkaline sludge.

3) Definition of the conditions for integrating all involved processes.

4) Patenting outfit-technological manners and solutions for accomplishment of the integrated technology.

5) Formulating the engineering standing order for the comparative technical-economical evaluation of the obtained data as applied to the tested coal and ash samples .

Cost and recoupment of the Demonstration Project creation . On the basis of the US Mines Bureau information /K. Griotheim and B. Welch, "Impact of Alternative Processes for Aluminum Production on Energy Requirements," JOURNAL OF METALS (USA) , Vol. 33, No 9, 1981, pp. 26-32/ and the ex-USSR Branch institute VAMI /Technical-Economical Report on Expediency of Processing Ekibastuz Coal Ash into Alumina, Cement and other Products, Leningrad, VAMI, 1979,198 p.; 1980,31 p./, also as applied to ash processing alumina plant with capacity of 100,000 tons alumina per year, the expected capital investment amount is $120 million. Capital investments for aluminum smelter with corresponding annual capacity about 50,000 tons aluminum are expected to be $250 million. So, the total capital investments for constructing coal ash-alumina-aluminum complex are $370 million. According to the website MetalPrices.com price of one ton of aluminum is $ US 1600. Consequently, marketing cost of 50,000 tons of alumina is $ 80 million. According to the material balance, nine tons of cement will be produced per one ton of alumina what corresponds to 900,000 tons of cement yearly. At cement price $168 per ton, the total annual sum of cement sales is expected to be equal $ 151.2 million. Thus, the total selling amount of both aluminum and cement is $ 231.2 million per year. On evidence delivered from Intec Engineering /*Aluminium Today Journal, v.9, 1997, p.44*/, profit of selling one ton of the commodity is estimated as 37.5%. Therefore, total annualy profit of the proposed Demonstrated Project is expected to reach $ 86.7 million. In this case, recoupment period is equal to 370:86.7=4.26 years. Moreover, new patents and know-how together with the Demonstration Project status will be additional very valuable product of the integrated complex and provide the further decreasing the recoupment duration of the Project errection.

5. THE EXTENT OF IPAAC COMPLETENESS

All parts of IPAAC, including boiler, units for crushing and grinding feedstocks, preparing slurry, ash digestion, separation of alkali-aluminate solution from sludge, solution desilication, decomposition and evaporation, furnaces and smelters for processing alumina into aluminum had been existed and operated at the Leningrad VAMI pilot plant during 1960s-1990s. At the present time, these facilities are dissambled and the pilot plant stops its existence.

5.1. **The designing and engineering documentations that are available:**
1)Designing documentations of the VAMI pilot plant has been certainly keeping in VAMI archive (St. Petersburg, Russia).

2)V.S. Sazhin, V.L. Rayzman, I.Z. Pevzner, et al., "Development of Scientific Basis and Engineering for Processing High-Silicon Bauxite and Other Aluminum-Containing Ores by the Combined Hydrochemical Method", Research Report Reg. (î 78076052, Leningrad-Kiev, VAMI-IONKh of Academy of Science of Ukrainian SSR, 1980.-180 p.

3)Technical-Economical Basis for Application of the Hydrochemical Method for Processing Nepheline Feedstock, Leningrad, VAMI, 1983.-131 p.

4)V.L. Rayzman, V.V. Volkov, L.P. Ni, R.A. Abdulvaliev, V.M. Pavlenko, et al., "To Develop in Pilot Plant Scale the Method of High-Temperature Hydrochemical Processing Low-Grade Aluminum-Containing Raw Material", Research Report, Reg. (î 01860025114, Leningrad, VAMI-IONKh of Academy of Sciences of the Ukranian USSR -IMIO of Academy of Sciences of the KazakhSSR-Gidrotsvetmet, 1990.-148 p.

5.2. The level to which the integrated technology has been developed

All parts of the integrated process are either industrially applied (steam and power generation, sodium aluminate production, aluminum hydroxide precipitation and calcinations, electrolysis of alumina-cryolite melt) or were pilot plant tested at VAMI plant in Russia /*V.S. Sazhin, "New Hydrochemical Methods for Completed Processing Aluminosilicates and High-Silicon Bauxite", Moscow: Metallurgiya, 1988, 213 p.*/, Alcoa plant in the USA /*H.J. Hittner, "Hydrothermal Alkaline Process to Extract Alumina from Anorthosite," TRAV. COM.INT. ETUDE BAUXITES, ALUMINE, ALUM., Paris, France, v. 16, 1981, pp.13-21*/ and ZSNP unit in Slovakia /*I.Podushkova and V. Novotna , "Production of Al2O3 from Nonbauxite Raw Material by the Hydrometallurgical Method", Budapest: Bulletin SEV, No 12, 1979, pp. 157-165*/ (lime-alkali digestion of aluminosilicate, separation of digested slurry into AAS and alkaline sludge, recovering alkali from that sludge, and AAS evaporation). Practicability of all stages of the integrated technology has long been proved just as industrially (thermal power plants, alumina refineries and aluminum smelters), so in pilot-scale (the hydrothermal method) and beyond question. The objective of the proposed project is adjustment (fitting) of the above-mentioned well-known and assimilated manufactures.

6. PROTECTION of the INTELLECTUAL PROPERTY

5.1.List of the priority papers

Inventor's certificates and patents concerned with the hydrothermal technology and being suitable for applying to the integrated method are listed in Table 2. We don't have any information about maintaince of these papers possessed by the VAMI as a patentee.

5.2. List of the published papers related to the proposed technology

Principle papers are listed in the above-mentioned References.

5.3. Topic of supposed "know-how"without its revealing

Know-how is concerned to coal combustion, ash digestion, conversion of AAS to aluminate solution, decomposition of the obtained solution, alkali regeneration, and combination of the production procedures.

5.4. Investigation of patent prototypes

Such investigation has been conducted as applied to patents listed in Table 2 and covered 15-20-years period in the following countries: USA, UK, France, ex-USSR, Germany, Italy, Australia, Greece, Turkey, India, Japan, and Hungary.

5.5. Organizations that finance research and design of the development
Ministry of Nonferrous Metallurgy of the former USSR had financed all research and design works during 1960s-1990s. Further developments was carried out by the authors without financing.

5.6. Country where the development started

Development started at Leningrad VAMI Pilot Plant (St.Petersburg, Russia).

Table 2 Patent documentation related to the proposed project Inventor's agreements of the ex-USSR (patentee - VAMI)

Title	No	Published in Bulletin of Inventions: year, No, p.
Method for drying slurries and solutions of alumina production	633238	1978, 42, 189
Method of sodium hydroaluminate production	778152	1980, 41, 272
Method for processing low-grade aluminum-containing raw material	820164	1981, 13, 237
Astringent	864735	1981, 34, 307
Method of silica brick production	1028622	1983, 26, 67
Method of alkali regeneration	1031089	1983, 27, 237
Method of sodium aluminate production	1032717	1983, 28, 227
Method for cleaning heat exchangers	1153652	1985, 16, 187
Method for processing aluminosilicate raw material	1220260	1986, 11, 257
Method for processing high-silicon raw material	1295666	1987, 5, 286
Method of aluminate production	1464426	
Method for processing aluminum- containing raw material	1508530	
Method for hydrochemical processing aluminosilicate raw material	1522660	
Method for recovering alkali	1537646	

Table 2 Patent documentation related to the proposed project Inventor's agreements of the ex-USSR (patentee - VAMI) (continuation)

Patent "Method of production of alkaline metal hydroaluminate"

Country	No	Date	Country	No	Date
Greece	852056	13 Jan 1986	Turkey	22725	24 May 1988
France	2609706	19 May1989	Australia	575223	21 July 1988
India	162818	20 Jan 1989			

On behalf and on the instructions of the authors:

Viktor L. Rayzman, Doctor of Technical Sciences

Tel: (310) 451-1028 E-mail: viktorayzman@yahoo.com and viktorayzman@verizon.net

Content

Preface...2

Introduction 3

Chapter 1. DEVELOPMENT OF ALKALINE METHODS

FOR PROCESSING RED MUD AND HIGH-SILICA BAUXITE...................6

1.1.Chemical Beneficiation of Red Mud...6

    Scandium Production, Value and Use...6

    Testing the Theory...7

    Advantages of the Chemical Beneficiation Method.............................10

1.2.Consecutive Recovering Silica and Alumina from High-Silica Bauxite

    with Different Fe_2O_3 Percentages...12

    Conversion of Minerals in Bauxite Conditioning..............................13

    Processing Bauxite 14

    Utilization of Alkali-Silicate Solution...18

    Conclusion..20

Chapter 2. UTILIZATION OF ALUMINA-BEARING COAL ASH...................21

2.1.Review of the Methods Dedicated for Producing Alumina

    and Byproducts from Coal Ash..21

    Coal Ash Composition...21

    Coal Desulfurization..23

    Integrated Technology for Coal Combustion and Ash Sintering...........24

    The Nucla CFB Clean Coal Technology.......................................27

    Acidic Leaching Coal Waste...... 28

    Alkaline Processing Coal Ash...... 34

    Sintering Method for the Processing Coal Ash into Aluminum

    and Silicon-Containing Products...34

Hydrothermal Alkaline Processing of Coal Ash...............................36

Alkaline Pre-Extraction of Silica from Coal Ash.............................38

Pilot Plant Tests and Prospective Technology for Processing

Coal Ash and other Nonbauxite Raw Materials into Alumina

and Byproducts..39

Discussion Concerning the Reviewed Processes............................42

2.2. Utilization of Lignite Ash by Alkaline Hydrothermal Method.............50

The Ways for ALHTP Improvement..55

Discussion and Conclusion...57

2.3. Heat Exchanger Fouling in Alumina Refinery

as a Mass Transfer Phenomenon...59

Kinetics of Heat Exchanger Fouling in Alumina Manufacture............59

Pilot Plant Test of Theoretical Predictions....................................61

Application of the Test Results..63

The Most Efficient Way for Mitigating Heat Exchanger Fouling..........63

2.4. Regularity of Hydrothermal Chemical Reactions.........................65

2.4.1. Development..66

2.4.2. Results...68

2.4.3. Discussion..72

2.4.4. Conclusion..74

Chapter 3. INNOVATIONS in PROCESSING LIQUID ALUMINATE

INTERMEDIATES and WASTES..77

3.1. Effect of Deep Aluminate Solutions Desilication

on Alumina Yield in the Precipitation Procedure......................77

Silicon State and Behavior in Aluminate Solution....................77

Impact of SiO_2 Concentration on the Precipitation

of Alumina Hydroxide...78

145

Desilication of Aluminate Solution Produced by Leaching

of Sintered Material..79

Feasibility of Applying Calcium-Containing Precipitants

for More Complete Desilication of Aluminate Solutios

in the Bayer Process..81

Conclusion...81

3.2. Extracting Sodium Aluminate from Aluminate Solitions

and Liquid Wastes..83

Sodium Aluminate Use..83

Equilibrium and Transitions in the Na_2O-Al_2O_3-H_2O

System..85

Processing Industrial Intermediate Materials into Sodium

Aluminate..89

SHA Precipitation from Aluminate Solutions in the Bayer

Process..89

SHA Precipitation as an Application to the Sintering

Process..89

SHA with Reference to the Hydrothermal

Process..91

Sodium Aluminate Recovery from Industrial Liquid

Wastes...91

Sodium Aluminate Manufacturing from Solid Wastes............94

Conclusion...99

3.3. Stability of Alkali-Aluminate-Silicate Solutions in the

Na_2O-Al_2O_3-SiO_2-H_2O System.....................................104

Chapter 4. INTEGRATING COAL COMBUSTION AND PROCESSES

FOR RECOVERING ALUMINA FROM INDUSTRIAL WASTE....108

146

4.1.About Co-processing Coal Ash and Red Mud into Alumina 108

    Conclusion... 109

4.2.Creation of Power-Chemical-Metallurgical Complexes as

    the Way to Solve Both Raw Material and Fuel Problems

    in Aluminum Industry..112

    Technical-Economical Comparison of the Alkaline

    Methods Used for Processing Aluminosilicates......................112

    Integrated Technology as Applied to Different Types

    Of Alumina-Bearing Raw Material....................................117

    Some Directions for Modification of the Integrated Technology...125

    Conclusion...126

4.3.Thermochemical Evaluation of Industrial Methods for

    Processing Alumina-Bearing Raw Materials and Waste...........128

ATTACHMENT. Proposals on a Project "New Technology for Integrating

    Production of Power, Chemicals, and Aluminum"...................133

Content...144

THE AUTHOR RAYZMAN VIKTOR LAZAREVICH

was born in Russia in 1934. From 1944 to1991 lived in Leningrad (St. Petersburg). Earned Metallurgical Engineer (1957), Ph.D. (1978), and Doctor of Technical Sciences(1990). Immigrated to the United States In 1991. Has been granted 32 patents. The author or co-author of 8 books and 115 technical papers published in

Russia, Ukraine, Kazakhstan, USA, England, and Thailand.

List of last published papers (in English):

-Technologies of Coal Fly Ash Processing into Metallurgical and Silicate Chemical Products (Report). *Proceedings of the 210th American Chemical Society National Meeting: Chicago, IL, August 20-25, 1995, p.863-867. Co-authors: S.A. Shcherban and I.Z. Pevzner.*

-Alkaline Terchnologies for Coal Fly Ash Processing into Metallurgical and Silicate Chemical Products (Poster).*Proceedings of the 1995 International Ash Utilization Symposium (University of Kentucky and the Journal FUEL): Lexington, Kentucky, October 23-25, 1995. Co-author: S.A. Shcherban.*

-Mass Transfer Considerations During the Precipitation of Deposits on Alumina Refinery Heat Exchanger Surfaces (Report). *Light Metals 1996 [Proc. Conf], Anaheim, CA, USA, 4-8 February 1996 (TMS/AIME: Warrendale, PA,1996),p.5-10.*

-More Complete desilication of Aluminate Solution is the Key-Factor to Radical Improvement of Alumina Refining (Report), *Light Metals 1996 [Proc. Conf], Anaheim, CA, USA, 4-8 February 1996 (TMS/AIME: Warrendale, PA,1996),p.109-114.*

-Integrated Technology for Combustion of Aluminum-Containing Coal and the Recovery of Alumina and Byproducts from the Formed Ash (Report), *Light Metals 1996 [Proc. Conf], Anaheim, CA, USA, 4-8 February 1996 (TMS/AIME: Warrendale, PA,1996),p.151-157.*

-Recovering Alumina, Silica and Byproducts from Coal Ash Through the Use of Process for Silicon Pre-Extraction (Report), *Light Metals 1997 [Proc. Conf], Orlando, Fla, USA, 9-13 February 1997 (TMS/AIME: Warrendale, PA,1997),p.133-136. Co-author S.A. Shcherban.*

-Technology for Chemical-Metallurgical Coal Ash Utilization (Review), *ENERGY&FUEL, v.011, No 004, 1997, p.761-773. Co-authors: S.A. Shcherban and R.S. Dworkin.*

-Coal Ash Utilization by the New Chemical-Metallurgical Processes,*Proceedings of 1997. Co-author: S.A. Shcherban.*

-Red Mud Revisited – Special Paper on Scandium Potential, *ALUMINIUM TODAY, 1998, v.10, No 5, p.64-68.*

-Sodium Aluminate from Alumina-Bearing Intermediates and Wastes (Overview), *JOURNAL OF METALS, November 1998, v.50, No 11, p.32-38. Co-authors: I. Filipovich, L. Nisse, Yu. Vlasenko.*

-Integrating Coal Combustion *and Red Mud Sintering at an Alumina Refinery, JOURNAL OF METALS, v.51, No 8, 1999, p.16-18. Co-author: I.K. Filipovich.*

-Thailand's Lignite Ash as a Raw Material for Manufacturing Alumina and Byproducts by Alkaline Hydrothermal Method, *SURANAREE JOURNAL OF SCIENCE AND TECHNOLOGY, v.8, No 1-2, 2001, p.83-89. Co-authors: P.Rerkjirasawad, A.Aturin.*

-Extracting Silica and Alumina from Low-Grade Bauxite,*JOURNAL OF METALS, No 8, 2003, p.45-48. Co-authors: I.Z. Pevzner, V.M. Sizyakov, L.P.Ni, I.K. Filipovich.*

List of the published books (in Russian):

-Application of Hydrothermal Processes for Digestion of High-Silica Aluminum-Containing Raw Materials,. *TSVETMETINFORMATSIYA, Moscow: 1981, 40 p. Co-authors: M.M. Neusikhin, Yu.K. Vlasenko and I.Z. Pevzner.*

-Autoclave Processes in Alumina Production, *METALLURGY, Moscow: 1983, 128 p. Co-author: I.Z. Pevzner.*

-The Compatible Processing Bauxite and Nepheline for Production of Alumina, *TSVETMETINFORMATSIYA, Moscow:1984, 52 p.*

-Production and Application of Sodium Aluminate, *TSVETMETINFORMATSIYA, Moscow: 1987, 48 p. Co-authors: Yu.K. Vlasenko, L.S. Nisse and L.S. Sinelshchikova.*

-Chemical Beneficiation of High-Silica Aluminum-Containing Raw Materials, *TSVETMETINFORMATSIYA, Moscow: 1987, 60 p.Co-authors: L.P. Ni, N.S. Malts, S.A. Shcherban, A.N. Naumchik and V.M. Pavlenko.*

-Combining Methods of Processing Low-Grade Aluminum-Containing Raw Materials, , *Alma-Ata: 1988, 256 p. Co-author: L.P. Ni.*

-Sodium Aluminate and Sodium Hydroaluminate. Production and Application, *ROSTOV STATE UNIVERSITY, Rostov: 1991, 120 p. Co-authors: Yu.K. Vlasenko, L.S. Nisse, A.P. Bogdanov, et al.*

-Alumina Production. Reference Book, *IMIO MN-ANRK, Almaty (Kazakhstan): 1998, 356 p. Co-authors: L.P. Ni and O.B. Khalyapina.*

www.ingramcontent.com/pod-product-compliance
Lightning Source LLC
Chambersburg PA
CBHW080253180526
45167CB00006B/2520